THE COMPLETE GUIDE TO
WATER STORAGE

How to Use Gray Water and
Rainwater Systems, Rain
Barrels, Tanks, and Other
Water Storage Techniques for
Household and Emergency Use

By Julie

The Complete Guide to Water Storage:

How to Use Gray Water and Rainwater Systems, Rain Barrels, Tanks, and Other Water Storage Techniques for Household and Emergency Use

Copyright © 2012 by Atlantic Publishing Group, Inc.
1405 SW 6th Ave. • Ocala, FL 34471 • Ph: 800-814-1132 • Fax: 352-622-1875
Website: www.atlantic-pub.com • Email: sales@atlantic-pub.com
SAN Number: 268-1250

Library of Congress Cataloging-in-Publication Data

Fryer, Julie.
   The complete guide to water storage : how to use tanks, ponds, and other water storage for household and emergency use / by Julie Fryer.
      p. cm.
Includes bibliographical references and index.
ISBN-13: 978-1-60138-363-1 (alk. paper)
ISBN-10: 1-60138-363-0 (alk. paper)
   1. Water reuse. 2. Water--Storage. 3. Water harvesting. 4. Graywater (Domestic waste water)--Recycling. I. Title.
TD429.F79 2011
628.13--dc23

                        2011030933

Printed in the United States

PROJECT MANAGER: Gretchen Pressley • gpressley@atlantic-pub.com
BOOK PRODUCTION DESIGN: T.L. Price • design@tlpricefreelance.com
PROOFREADER: Sarah Wilson • sarah.alisha.wilson@gmail.com
FRONT COVER DESIGN: Meg Buchner • megadesn@mchsi.com
BACK COVER DESIGN: Jackie Miller • millerjackiej@gmail.com

Printed on Recycled Paper

A few years back, we lost our beloved pet dog Bear, who was our best and dearest friend and the "Vice President of Sunshine" here at Atlantic Publishing. He did not receive a salary but worked tirelessly 24 hours a day to please his parents.

Bear was a rescue dog who turned around and showered me, my wife, Sherri, his grandparents Jean, Bob, and Nancy, and every person and animal he met (well, maybe not rabbits) with friendship and love. He made a lot of people smile every day.

We wanted you to know a portion of the profits of this book will be donated in Bear's memory to local animal shelters, parks, conservation organizations, and other individuals and nonprofit organizations in need of assistance.

– *Douglas & Sherri Brown*

PS: We have since adopted two more rescue dogs: first Scout, and the following year, Ginger. They were both mixed golden retrievers who needed a home.

Want to help animals and the world? Here are a dozen easy suggestions you and your family can implement today:

- *Adopt and rescue a pet from a local shelter.*
- *Support local and no-kill animal shelters.*
- *Plant a tree to honor someone you love.*
- *Be a developer — put up some birdhouses.*
- *Buy live, potted Christmas trees, and replant them.*
- *Make sure you spend time with your animals each day.*
- *Save natural resources by recycling and buying recycled products.*
- *Drink tap water, or filter your own water at home.*
- *Whenever possible, limit your use of or do not use pesticides.*
- *If you eat seafood, make sustainable choices.*
- *Support your local farmers' market.*
- *Get outside. Visit a park, volunteer, walk your dog, or ride your bike.*

Five years ago, Atlantic Publishing signed the Green Press Initiative. These guidelines promote environmentally friendly practices, such as using recycled stock and vegetable-based inks, avoiding waste, choosing energy-efficient resources, and promoting a no-pulping policy. We now use 100-percent recycled stock on all our books. The results: in one year, switching to post-consumer recycled stock saved 24 mature trees, 5,000 gallons of water, the equivalent of the total energy used for one home in a year, and the equivalent of the greenhouse gases from one car driven for a year.

Table of Contents

Chapter 2: Common Sources of Water 67

Chapter 3: Ways To Store Water 91

Chapter 4: Installing a Gray Water System 117

Chapter 5: Installing a Rainwater Harvesting System 153

Chapter 6: Water Quality 189

Chapter 7: Keeping Your Storage Systems Running Smoothly....................... 213

Chapter 8: Using Your Water 227

Appendix B: Formulas and Conversion Tables 291

Appendix C: Assessment Form 295

Introduction

Water — that magical, flowing substance we see and use every day. We use it in our bathtubs, our lawns, our sinks, and even our cars. We drink it, cook with it, and wash with it. Our planet is more than 70-percent water, so it is hard to imagine a world in which we will not have as much fresh water as we need to do our laundry and quench our thirst. However, less than 1 percent of that water is usable. Accidents and disasters happen that can render even the most affluent or water-prosperous consumer without water for an unknown time.

Most people do not think about water storage or what would happen if there were a serious water shortage. It might seem like a strange idea to even consider drinking rainwater or recycling gray water (the waste water from washing machines and showers) to water your yard. If you live in a place where rainwater is abundant, it might not even occur to you to conserve that valuable rainwater for your yard, but as you will learn, water conservation is a global issue. Even if you do not need extra water for household use, it is always an idea to consider some emergency water storage.

A well-known example of the need for emergency water is to review what happened in 2005 when Hurricane Katrina struck South Florida and the Gulf Coast. Even with days of advanced warning, the aftermath of this storm left millions of people without access to clean water. Even with the mobilization of emergency crews and relief organizations handing out bottled water, the demand was high. Long after the hurricane hit, there was no electricity, and more important, there was no way to purify, pump, or supply clean water through municipal systems. Many people were unprepared and had no emergency water stores.

You also want to have water stored in case the local water system undergoes maintenance or an accident alters your access to water. Suppose a backhoe at a construction site punctures a water main, and your town is without water for a day or two. Suppose a routine water system check of the system finds bacteria, and the whole town must boil water for a few days before drinking it. There are numerous reasons to have a few days worth of clean drinking water on hand in your home. Do not take your water supply for granted. Events beyond your control can happen that would make the water stop coming out of your tap or that same water unsafe to drink.

What is Potable Water? Water that is "potable" is another term for water that is uncontaminated, clean, and safe for drinking. Non-potable water cannot be used for drinking, cooking, or food preparation.

Having readily available water at home in case of an emergency is much better than standing in line afterward and hoping they do not run out of water before your turn. A true emergency situation is stressful enough without worrying about where your drinkable water will come. Keep in mind, a person can survive without food for about three or four weeks, but you can only survive about a week without water. According to the Federal Emergency Management Agency (FEMA), a person needs at least a half a gallon per day to live, but it is recommended that a person store about a gallon of water a day for emergency situations.

There are many reasons people decide to store water for household or emergency use. For some, rain barrels or tanks can provide much-needed irrigation for gardens and lawns in areas with frequent droughts, while others use rain barrels to reduce their dependency on city water for irrigation.

Each person's needs are different. For instance, farmers may capture water in ponds or ditches to be used for drinking water for their livestock or irrigation needs. City dwellers might store water in 5-gallon containers in closets to be used in an emergency when water can be in short supply. Storing water can be as simple as filling a container from your kitchen tap or as complex as harvesting rainwater and passing it through several types of filters for use in the home.

Water Storage Around the World

▶ Here is an ancient cistern that still retains water at UNESCO's World Heritage Site of Sigiriya Lions Rock in Sri Lanka. This ancient rock fortress and palace ruin was built during the reign of King Kassapa I 477-495 AD.

Creating water collection and storage systems might seem like a new idea in many areas of the United States, but, in reality, these systems have been in place for hundreds of years throughout the world. Many rural areas and underdeveloped countries still rely on these methods as their only source of drinking water. From rainwater collection in cisterns to fog screens to deep wells, each region has adapted the many options for water capture and storage to meet their water needs. Some of these methods may be adaptable to your own situation, while others may be impractical for the region where you live. This book will explore how these systems have been modernized and adapted for just about any area. As you read this section, think about what you can do creatively to conserve in your home.

Rome

Ancient Rome was designed with an extensive aqueduct system to provide water for its residents. The rich also built homes with paved courtyards to direct water into rainwater-holding cisterns. Poorer residents would rely on public baths and water systems that were fed by the aqueducts.

Bermuda

There is no "city" water in Bermuda because the limestone base of the island makes installing water pipes throughout the island a difficult and expensive process. This means each individual household is responsible for its own water supply in many ways. Bermuda does not have any rivers or freshwater lakes and, thus, relies on rain for its water. Across the country, people use rainwater harvesting, store water in cisterns, and desalinate saltwater. These systems are used for everything from drinking water to flushing toilets. All new construction requires a rainwater harvesting system, and most residential buildings rely on gutters and cisterns to store household water. Government regulations require special whitewash roof paint to seal the roof and reduce the amount of contaminants in the water. Cisterns or water tanks must also be painted on the inside to keep contaminants from leeching into the water.

Thailand

▶ Rainwater jars in Thailand.

Although Thailand has a long monsoon season when water is plentiful, they also experience a dry season that lasts up to four months. To store up water for these times, many villages in Thailand use rain jars to store rainwater. These jars can hold up to about 2,000 liters (about 530 gallons). They are made out of clay or concrete and have tight-fitting tops when it is not raining. The jars keep the water safe from mosquito infestation, which can lead to malaria and other illnesses, and from becoming dirty and unsuitable for drinking. The tradition of using the jars is 2,000 or more years old but was revived in the 1980s as a water solution for rural areas.

Australia

Australia is the driest continent on Earth. Rainwater harvesting and water storage for droughts are extremely important in the country. Reusing water, as in gray water systems, is common across the country. The Australian government requires that all newly constructed buildings have rainwater harvesting systems in place and provides many incentives and rebate programs for those who install rainwater harvesting systems in homes and businesses

Africa

Africa is the second driest continent on Earth, even though it contains Lake Victoria, the second largest lake in the world. Africa also has a massive land area, and not all of its countries share the same climate or access to water. Many of the wettest areas of Africa are also the least populated by humans, and, on average, most people have reduced access to clean water.

Most small villages, if not near a lake or river, rely on well water for all water needs. Many missionary groups and world health organizations work on well-building projects and frequently train local people on how to maintain and repair the wells so they can become self-sustaining. Many of today's most advanced purification systems have come out of the work done to help people in these regions.

Singapore

Singapore is a relatively urban country and one without an extensive natural water supply. Many residents of Singapore live in high-rise apartment buildings. The buildings contain rainwater harvesting systems on the roof that feed into large cisterns, also located on the roof. This is a source of gravity-fed, non-potable water for building residents, which saves both energy and potable-water resources. Water in the rooftop cisterns is reserved for toilet-flushing purposes and irrigation.

Brazil

Some areas of Brazil have semi-arid climates and have a need for additional water resources. Rainwater harvesting is fast becoming the solution to growing water needs in the semi-arid regions and the urban areas of Brazil. The idea of rainwater harvesting became so important in Brazil that in 1999 the Brazilian Rainwater Catchment Systems Association was formed. Several civil society groups started an initiative called the One Million Rainwater Harvesting Programme in an effort to provide reliable drinking water to rural residents of the semi-arid regions. According to the initiative, 1 million rainwater harvesting tanks will meet the drinking water needs of 5 million people. The tanks that are built hold an eight-month supply of rainwater — enough to get families through the driest time of the year. Small fish that eat larvae are kept in the tanks to keep the water quality high.

India

Tamil Nadu at the southern tip of the Indian Peninsula, is adept at collecting rainwater. This area relies heavily on monsoon rains to provide water; if the rains do not come as expected, droughts are common. During the rainy season, rainwater is harvested and as much water as possible is directed into the soil to recharge underground aquifers and wells. If aquifers and wells are recharged during the rainy season, people can survive the dry season without difficulty. The key is to make sure rainwater has the opportunity to slowly soak into the soil.

Canals, water tanks, seasonal rivers, and wells are the main focus of today's rainwater harvesting. Archaeologists have found water storage artifacts dating back almost 4,000 years, and many of these methods are still being used today.

Germany

▶ Here rainwater is collected and piped to a holding tank where it is used for irrigation.

Germany is a European leader in rainwater-harvesting programs. Most rainwater harvested in Germany is used in residential and commercial buildings for toilet flushing, washing machines, and landscape irrigation. The German Water Resources Act, which regulates water usage throughout Germany, calls for a safe and secure water supply for everyone. The act, which was first passed in 1957 and has been updated regularly since, also states that everyone should work to use water responsibly and economically and protect the available water resources. Most residential systems feed roof runoff into an underground cistern where the water is held for later use. Overflow is sent out through perforated pipes into the surrounding soil. About 50,000 systems are installed each year in Germany.

Water Storage in the United States

In America, we have access to a seemingly unlimited supply of usable water. According to the Environmental Protection Agency (EPA), the average family of four in the United States uses about 400 gallons of water a day. The United States consumes twice as much water as other industrialized nations and more water than less-developed countries. Although water is continually regenerated through the water cycle, we must find ways to conserve and create innovative designs to bring water to cities with ever-increasing populations.

During the 1800s, water cisterns and rainwater-capture systems were installed on many homes throughout the country. Even before settlers arrived, native populations used various means of collecting rainwater and using it for daily needs. As pioneers created homesteads and needed water for the families, crops, and livestock, they looked for ways to access water. Some were able to dig wells, and back in the 1800s, they had an easier time of it because water tables were much higher than they are today. Wind power or hand pumps pumped then these old wells, and water was brought into houses in a bucket. The arid climate areas had large stone cisterns to capture and hold rainwater for later use. Early towns were often formed near rivers or natural springs to have water easily available for residents.

As cities grew and these concentrated populations placed more demands on water supplies, the existing water tables shrunk. In addition, the waste and pollution from these sprawling metro areas contaminated the water available for use. Cities began developing municipal sewer systems that treated the water going to homes and removed the water that these homes used. Few of the new municipal systems worried about conservation or reuse of gray water. Over time, these systems spread, and people became used to clean drinking water at the lift of the faucet handle. This readily available water made it possible to develop areas, such as Las Vegas and southern California, where it would normally have been difficult to sustain life because water was so scarce.

As municipal water supply became more prevalent throughout the United States, conservation and water collection took a back seat. Collection water from the roof or reusing gray water seemed like a practice from the olden days, and people moved away from the methods of water collection. Now that we see the need for water conservation, these old-time remedies for water shortages are coming back into style. Modern technology has also stepped in to upgrade these methods, so they are more applicable to current lifestyles.

Large populated areas also create a large area of covered surfaces that are nonpermeable. These shed rainwater quickly and create fast-moving runoff that leads to erosion. After some time, this erosion creates a path for the water to follow — such as a gully

— and the water will move even faster through the area. This fast movement impedes the soil from properly absorbing the rainfall and channeling it back into the water table. Without proper absorption, natural aquifers are not being replenished.

The current idea of rainwater harvesting in much of the United States involves placing a 55-gallon plastic barrel under a downspout to catch the rain. In desert areas throughout portions of Arizona and New Mexico and even some portions of Texas, rainwater harvesting is a few decades ahead of the suburban rain barrel, mostly out of necessity.

In Texas, at Hueco Tanks State Park outside El Paso, there are natural water cisterns. These basins are formed from the rock, and locals have used them for thousands of years. Native American artifacts found nearby are believed to be 10,000 years old, which indicates a long history of rainwater harvesting in the area. However, there are still hurdles for these systems to face. For instance, in North Carolina, it is not permitted to use gray water systems. The North Carolina Plumbing Code states that gray water is "waste discharged from lavatories, bathtubs, showers, clothes washers, and laundry sinks." The code goes on further to say that putting untreated water on plants, such as trees, flowers, and gardens, is illegal and unhealthy.

Everyone's situation will be different, and therefore, some of the suggestions in this book might need to be modified to work in your particular area. In addition, you will need to check with state and local agencies before digging or setting up any of the systems in this book because there are different regulations in each area that need to be followed. You will find suggestions of where you might find the same information in your particular state and municipality throughout the book.

What this Book Can Do for You

Water sustains life. That is a powerful reason to make sure you have at least some water stored, no matter where you live. Once you have drinking water, you can focus on storing water in your soil to recharge the land around you, which ensures your plants will be healthy and the aquifer beneath the soil will be full of pure, clean water whenever you might need it.

If you have never stored water and do not know of the many methods of doing so, this book will give you the information and tools you will need. Whether you live in an apartment, a suburban home, or a farmhouse on many acres in the country, you can store water for many uses, including watering plants or as an emergency drinking supply. Water collection and storage is a practice useful in every part of the country and for most any use. Even if you live in an area that has copious amounts of rainfall, learning to collect this rainwater and using it for the household will save water usage that normally would go through a treatment facility. If you live in an area where only a few inches fall each year, reusing your gray water will help you stretch your water usage without taxing your budget or well.

This book includes chapters about storing water in simple containers for emergencies, explains more complex water collection and storage systems, and provides basic ideas about capturing water for the benefit of the trees and plants on your property. It also includes simple water conservation tips you can use every day, what to do in a water emergency, and ideas about which water purification and conservation tools to have tucked away for an emergency. You will learn the types of water storage systems available, the costs of materials and installation, how to check your local regulations, and the ways you can use water once it is stored.

If you live in an area prone to droughts or water restrictions, this book will provide ways to store water in the soil around your home, which helps plants grow and flourish. Storing water in the soil can also help recharge the underground aquifers near you and raise the water table, which makes the land healthier. Even if you have a well, this book can teach you how to keep your well filled with clean, fresh water, even when the wells around you are going dry. You can even create a self-sustaining water cycle on your property if you plan ahead.

The most important thing you can do while you are reading this book is to take the time to observe the water around you to see where you are using the most water in your home and consider ways to use less or recycle some of that used water. For example, you can use a biodegradable soap in the laundry and then use the waste water to nourish your trees. If you live on a hill, go outside when it is raining and observe where the water is draining. Does it all run downhill, away from your land, and into the street? Are there natural areas where the water slows down and has time to soak into the soil? If much of the rainwater leaves your land quickly, you can slow the water flow down and keep it on your land for as long as possible

As you begin to think about water storage, consider the origin of your water. Do you have a well, or are you on city water? If you have city water, where does it come from? Taking the time to think about and observe the water around you will give you some ideas about how to save and store the water available, thus ensuring you and your family will have safe, fresh water no matter what circumstances may come your way.

Every thorough discussion of storing water begins with collecting the water and ends with how to efficiently use this cache. To adequately prepare yourself and put your water to use, you will need to know about the many ways possible to collect water — even if you live in a desert, there are methods for gathering enough water. You also must know how to treat this water so it is safe for your family, livestock, pets, and garden. You will have to learn how to store this water so it stays clean and is accessible even in the most trying times. This book will cover all those topics and offer some concrete suggestions on how to set up and use your own rainwater or gray water collection system and offer ideas on best ways to use this water, whether it is for watering a tree or holding your family through a natural disaster.

Why Should You Collect or Store Water?

There are two main reasons to collect or store water: for emergencies and for everyday use. In some instances, one system can take care of both needs, or multiple systems can be combined. The design options are unlimited and can be adapted to meet any specific need. These larger systems require a bit more upfront engineering and ongoing maintenance, but many large businesses, schools, and municipal entities are now implementing their own collection and storage plans. At home and on a large corporate scale, these systems reduce utility bills and usage, efficiently use water resources, and often allow businesses to garner environmental awards. Water-conscious construction or remodeling can help a homeowner or business achieve LEED certification and valuable tax breaks.

What is LEED certification? LEED stands for Leadership in Energy & Environmental Design. The U.S. Green Building Council developed the building verification system, which certifies green building practices. LEED certification is recognized around the world as the highest achievement in environmentally sound construction. The certification gives your project the stamp of approval for potential buyers and allows participation in many government incentive programs. There are numerous specific requirements and varying levels of certification — water conservation, however, ranks high in importance. To fully understand the program and learn more about the requirements, visit the Council's website at **www.usgbc.org** for more information. If one of your project goals is to become LEED-certified, it is important to plan from the beginning and consult with a professional versed in LEED-building practices.

This chapter will talk about the benefits of storing and collecting water for all purposes and offer some basics on storing water. It will help you determine how much water your family might need and provide information on assessing water needs for your family in the event of a disaster. It will also cover household use, which includes anything from watering plants to setting up a system to harvesting rainwater. Focus on emergency water storage for now, but if you decide to do more, you will have plenty of information to get started.

Water Statistics

On the planet, all the water there will ever be is already there. It is recycled constantly, but all there ever was is in the air or on the land and makes up the oceans. Seventy percent of the surface of the planet is covered in water, but less than 1 percent is usable. Consider the following facts:

- Only 2.5 percent of the water on the planet is fresh water. The other 97.5 percent is undrinkable salt water.

- Only 30 percent of the fresh water can be used. The other 70 percent is frozen in the polar caps in Antarctica and Greenland.

- Less than 1 percent of the fresh water not frozen is accessible for human use. The other 99 percent is in the soil or deep under the ground in aquifers.

This usable water is in rivers, lakes, reservoirs, and underground aquifers close enough to the surface.

Percentages of fresh water on Earth

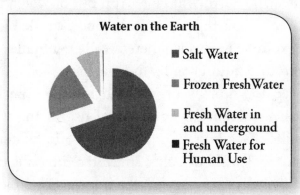

As the previous graphic shows, only a small percentage of the water on Earth is fit for human consumption.

The hydrologic cycle

To truly understand how to most effectively harness and conserve water, it is important to know how water moves through the atmosphere. Water can be found in three forms: solid (ice), liquid, or gas (water vapor). Water cycles through these forms in many ways, including: falling as rain or snow; soaking through the topsoil and into the aquifer; collecting in rivers, lakes, and oceans; and evaporating back into the atmosphere to start the process again.

In a basic sense, the water around when dinosaurs roamed the Earth is the same water humans drink now. This "movement" is referred to as the hydrologic cycle, and it means the atmosphere is always holding the water in the air, bodies of water, or the groundwater. The Earth will never run out of water, but human interference can make this water supply non-drinkable or hard to access. An example of this is when a well eventually "runs dry." The underground aquifer pulls the water available to this well, which is used too quickly or pumped too far from the source, such as with irrigation for agricultural use. If the water is not allowed to filter back into the aquifer, the groundwater source for the well is not replenished. Collecting and redirecting rainwater is only one of the ways people can step into this hydrological cycle without adversely affecting the regeneration of drinking water.

Hydrologic cycle

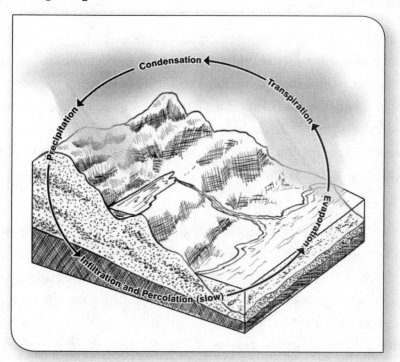

As you will see from the graphic, water evaporates in the atmosphere, condenses, and returns to the Earth as rain, sleet, snow, or hail. Water on the surface of the Earth evaporates, or turns into water vapor, and rises into the atmosphere. Plants also release water through a process of transpiration as part of their growth and life. Whether it is the water from an ocean or a freshwater lake, the process of evaporation is the same. Water transforms from a liquid into a gas through the process of evaporation and rises into the atmosphere. This water vapor cools and condenses to make clouds. Clouds eventually become saturated enough to release the water in liquid or frozen states.

Any unevaporated water that reaches Earth is water that can be used for human use.

After the water hits the soil, plant roots snatch up most of it. After the plant uses the water for its many metabolic processes, it releases it back into the atmosphere. If the water gets past plant roots, it can saturate the soil. The area in which this occurs is called the zone of saturation, and the water found in this zone is referred to as groundwater. The top part of the groundwater is referred to as the water table.

Water not saturated into the soil and not evaporated immediately back into the atmosphere is called surface water. Surface water can include water pooling on the ground and is water that runs off hard surfaces, such as rooftops, parking lots, sidewalks, and other manmade structures. Whether it goes directly into the ground, runs off a structure, or is filtered through a waste water treatment plant, this is the water that ends up in rivers, lakes, streams, and elsewhere. As you will learn, you can collect this surface water and direct water back into your groundwater supply through different techniques and equipment.

Water in the form of rain or other precipitation passes through pollution on its way to the ground. This is how issues, such as acid rain, occur. Due to pollution in the atmosphere, not all surface water is safe to drink. This is not to say rainwater cannot be used for other things, such as watering your garden, but it should not be used for drinking unless it is properly filtered.

Rainwater is mostly microbe-free. Microbes are microscopic creatures, such as yeasts, bacteria, and viruses. Rainwater only comes in contact with microbes after it reaches the ground or surface water. The water collected in your rain barrel is considered surface water, and it is not considered safe for drinking without further filtration or purification. Underground sources, such as springs where water comes in contact with the water table, feed some surface water. Some streams will run even during dry conditions because they are fed from an underground spring. During periods of drought, harvested rainwater can provide the necessary water for irrigating your yard and watering your plants.

Why Conserve Water?

In addition to storing your water, conserving water is essential, especially in areas prone to droughts or if you rely on water from a well. Even if you live in an area with water, being mindful of your water usage helps everyone by preserving the groundwater stores of water. Additionally, if you live off a municipal supply line, you pay for your water — and your city pays to treat, pump, and handle this water. By conserving the water you use for everyday household chores, you will save money and reduce the load on your municipality's treatment plant.

Water conservation is also becoming a necessity, so high-quality water will be available for future generations. The EPA predicts that at least 36 states will experience water shortages by 2013 at the local, regional, and/or state level. These water shortages are caused, in part, because the groundwater is pulling the water at a rate faster than it is returned to the aquifer. For example, housing developments in arid regions try to keep lush green lawns alive by frequent watering. These areas have difficulty supporting this type of non-native landscaping, and large amounts of water end up being used for non-drinking purposes and non-essential plant growth. This puts undue strain on the water source — whether it is a deep, on-site well or a municipal source, the water originates in the underground aquifer. The natural "recharge" cycle is disrupted, because water is drawn out more quickly than it enters the aquifer, and this leads to a shortage of water. By collecting rainwater instead of using tap water for landscaping, you can direct the water to a better use, avoid using groundwater unnecessarily, and reduce the inflow into your city's municipal system.

Each person is responsible for the world's water resources, and the way he or she uses water today can have an impact on the amount of water he or she has to use in the future. After you learn more about efficiently using water, such as watering your lawn with gray water, you will see conservation is quite simple. Even on a small household level, significant quantities of water can be used more efficiently and conserved with little extra effort.

Water Storage Benefits

Whether you are harvesting rainwater for use in your home, recycling gray water for irrigation in your yard, or reusing bathroom sink water to flush the toilet, there are environmental and financial benefits to finding and storing alternative sources of water for household use.

Environmental benefits

If you have a variety of water-saving and storage systems in place in your home, you can make the best use of the water available to you. For example, in an ideal setup, rainwater harvested from the roof can be filtered and used for most household purposes. *These rainwater catchments will be discussed further in Chapter 5.* Much of it can be captured and reused for flushing the toilet and watering the garden. So instead of running off your roof into the gutter and straight to your waste water treatment plant, this rainwater can be directed to your garden, where it will water your plants, naturally filter down through the soil and go back into the aquifer as groundwater supply. That aquifer will be responsible for keeping your personal well full, or it might be needed to feed the city water supply. Either way, it is a much more efficient and water-conserving approach than relying on the local city to supply water to your tap and deal with whatever flows down the drain.

▶ Water Treatment Facility

Capturing water from alternative sources reduces the demand for chemically treated municipal water. Using less city water means the city's water treatment facility will last longer, there will be less waste water entering the system, and the city will be able to use fewer chemicals, such as chlorine, sodium, or fluoride, to treat water. What happens to waste water after it enters the treatment system varies by region and depends greatly on climate, codes, and availability of water. Some areas try to reclaim water, while others treat it and release it into a drainfield. Either way, the less water sent to the treatment plants and the fewer resources used to treat it, the sooner people have access to clean water again.

Financial benefits

Another benefit to using alternative water sources, such as rain harvesting systems, is the reduced utility cost. If you are on a city-supplied water system, you pay for the water you bring into your house, the waste water you send out, or a combination of the two. Harvesting rainwater, even just for simple lawn and garden use, can significantly reduce your monthly water bill. This can also be said for reusing gray water; you get two uses out of the water you paid for, which reduces the cost for your household or business.

In the average American home, about 70 percent of water is used inside the house, according to the EPA. The remaining 30 percent is used outside for activities, such as watering the lawn or flower garden. Gray water is considered safe on ornamental plants, and rainwater is better than city water for plants because it is chlorine free.

Even if you only make changes to the water you use for irrigation, saving 30 percent of your water will reduce your water bill. What will your savings be? That depends on the price of your water. Rates vary around the country, from $23 a month in Memphis to $99 a month in Seattle for the average customer, according to a 2009 Austin Water Utility survey of 30 cities.

Local water districts charge anywhere from less than 1 cent to 10 or 15 cents per gallon of water. Examine your water bill to find out the rate you are paying. When reading your water bill, you might not be familiar with the terminology used. Some municipalities bill you by the number of gallons used. Others will charge by 100 cubic feet, or CCF. A CCF of water is equal to 748 gallons. You might pay only a few dollars per CCF, but it still adds up. Use the following equation to determine how much you are paying for water per gallon and how much you can save if you harvest rainwater for irrigation:

If you pay by CCF (based on rates in Erie County, Ohio, Summer 2010):

- Number of CCFs × Price per CCF = Amount you spent that month
- Number of CCFs × 748 = Number of gallons you used that month

- Amount spent ÷ Number of gallons = Price per gallon
- Amount spent × 0.3 = Amount saved by reducing usage 30 percent
- Example: Your bill said you used 11 CCF, and your rate is $3.90 per CCF.
  - 11 CCFs × $3.90 = $42.90
  - 11 CCFs × 748 = 8,228 gallons used
  - $42.90 ÷ 8,228 = $0.005 per gallon
  - $42.90 × 0.3 = $12.87 saved per month

Reducing your water bill by 30 percent will save you $12.87 per month. Although this might not seem like much money, by the end of a year, you will have saved $154.

If you pay by the gallon (based on rates in Collin County, Texas):

- Number of gallons × Price per gallon = Amount you spent that month
- Amount spent × 0.3 = Amount saved by reducing usage 30 percent
- Example: Your bill said you used 10,500 gallons, and your rate is $0.003 per gallon
  - 10,500 gallons × $0.003 = $31.50
  - $31.50 × 0.3 = $9.45 saved per month

Water prices across the country vary widely, so look at your bill to determine what your savings will be. Be aware some cities have base prices with additional charges if you exceed a certain amount of gallons, such as a price increase after you reach 35,000 gallons.

The EPA estimates, on average, a family of four uses 400 gallons a day. This totals 12,000 gallons a month. Because this figure is an average, it means families consume less and more than that amount. Consider the difference between a suburban home and an apartment; a suburban family might fill up the pool, water the landscape, let the kids run in the sprinkler, and wash cars. It would not be out of line for a family of four in a suburban home to see a summer water bill for 30,000 gallons of water or more. If your local water district charged you a half-cent per gallon, that would come to $150. Saving only 30 percent from that bill would save you $45.

Although 30 percent is the average amount of water used for irrigation, this number includes people who never water the lawn, who live in rainy areas, and who might only use a 1-gallon watering can once per week for irrigation. If you have built-in lawn sprinklers and use them at least once a week, a vegetable or flower garden, or young trees that require frequent waterings, your irrigation might count for about 50 percent of your water bill each month, instead of 30 percent. If you replace water with harvested rainwater, this will cut your water bill in half.

CASE STUDY: RAINWATER HARVESTING LESSONS FROM TEXAS

Pablo Solomon
Artist & Designer
musee-solomon@earthlink.net
www.pablosolomon.com
(512) 564-1012

I have been involved in environmental issues, conservation, and environmental design for most of my life. Growing up in Texas, I have seen the importance of water storage. I have also seen the many varied ways people have stored water here for hundreds of years. I like being independent and self-reliant. I think many other people do as well. I also like to think I am a steward of the Earth.

The United States is on par with other countries when it comes to rainwater harvesting and alternative water-use systems — we do as little as possible until necessary. Many places in the United States do the job because they have to, and because it is economically feasible. Where there is plenty of water, people tend to waste water.

The real challenge for the future is to desalinate seawater economically. It is a must-do. We also must get better at distributing water from areas of abundance to areas of need. Filtration to make water safe for drinking must improve and become available around the world.

▶ A low, small dam was created to save water for later use during dry periods. (Pablo Solomon)

My wife and I save all the water we can. We save rainwater in all types of containers — from a container holding hundreds of gallons, to used, plastic, gallon-size containers. We also have put small dams on our creek, which allow the wildlife to maintain normal movement, yet hold back water for dry spells. We have an 8,000-gallon concrete tank to store clean water pumped from our spring. This gives us clean water for household use.

Water is a precious commodity. In our area, one year can be flood after flood, followed by years of drought. We also store water in case of emergencies, such as tornados and ice storms, in which normal electricity and piped water can be disrupted. On many occasions, we have had to depend on the water and water pressure from our concrete tank after the electricity has gone out. We have enough water to last us a month or more.

▸ Pablo's cat, Hondo, visiting one of his 700-gallon metal rain collection tanks (Pablo Solomon)

In my opinion, the benefits of using stored water for things, such as flushing toilets and watering plants, just make sense. When setting up a water catchment system, you must be aware of mosquito control, prevention of contamination, avoidance of leaks, and overflow that damages your home or property.

I believe all water you use for drinking needs to be tested first. Then, you need to filter and sanitize the water, if necessary. Most places require any water served to the public to meet standards and be tested regularly. For example, if you own a restaurant or bed and breakfast, you must have an approved water supply.

Everyone should consider storing rainwater. Whether on a farm, in a neighborhood, or in an apartment, you can save water. Do what you can afford and what makes sense for your situation.

Storing Water for Emergency Use

FEMA recommends households have at least a three-day emergency water supply. The agency recommends a gallon per person per day. Half that gallon will be used for drinking, which equals eight glasses of water — the amount a person should drink daily for health. If you live in a warm climate or have other circumstances, such as medical conditions that warrant more drinking water, you will need to take that into consideration when you plan your emergency water supply.

The other half gallon of water is for cooking and cleaning. You will need to make some personal choices about your bathing options. If you are comfortable with sponge baths and are careful with your water usage, half a gallon might be enough. You also need to consider the types of food you will be eating in an emergency. If the foods require plenty of water for cooking, such as beans, rice, or pasta, you must address those needs in your water storage plan. To find more about FEMA's recommendations, you can download a pamphlet at **www.fema.gov/pdf/library/f&web.pdf.**

Many survivalist groups and household emergency planning experts recommend having one to two weeks' worth of water on hand to be well-prepared. The following sections will give you some idea of how much water your household will need each day and how to determine the number of day's worth of water you want to store. Many factors, such as space and geographic location, will influence your water storage plan. There are no quick rules for emergency water storage beyond the recommended gallon per person per day, but there are plenty of considerations when figuring out your own emergency water plan. The following sections list some common concerns and will get you thinking about what will work for your own specific situation.

How much water will you need?

First, you must consider the size and needs of your family. Think about daily life at your house and how much water is used. Showering, drinking, cooking, washing, and flushing the toilet require water. If your water supply is completely cut off, you might be willing to shower less frequently, and you can make choices about using the toilet, depending on your living situation. What you cannot ignore is having drinking water and water for cooking.

Your pets also will need drinking water. Dogs require about an ounce of water for each pound of body weight per day. There are eight ounces in a cup, so a small dog might only need two cups of water, but a large, 100-pound dog will need 12 ½ cups of water. Use the following formula:

Dog's weight in pounds ÷ 8 = Water needed in cups.

For more help with liquid conversions, see Appendix B.

Cats need less water than dogs, but you can use this same formula to determine how much water your cat will need. Just like people, your pets might require additional water if you live in a hot climate or they have special medical needs.

Spend a day or two making notes on the water you use throughout the day. If you can, try to measure the actual amounts. Does it take 4 gallons to wash the dishes, or can you do it with 1 gallon? Can you shower in five minutes or less if necessary? Make sure to include everything, including brushing teeth, making coffee, hand washing, and watering a vegetable garden if it is an important source of food for your family. *Chapter 8 includes extensive suggestions for rationing and efficient emergency water usage.*

Calculating Your Water Use

Try this exercise to determine how much water you use in a day:

1. Place empty gallon jugs near each water source in your house, such as the bathrooms, kitchen, and laundry room, and put a permanent marker with each jug.

2. For an entire day, use only water from the jugs. Fill up the jug with water from your tap and make a tally mark on the jug with your marker. Use water from the jug in every part of your daily routine, from brushing your teeth to cleaning your dishes.

3. If you use up a jug, refill it and put another tally mark on the jar. If you do not want to mark your jugs, keep a note pad near them to record how often you refill the jugs.

4. It might be more difficult to determine water needs for larger tasks, such as washing clothes or showering. Check your instruction manual or manufacturer's website for a guideline on water usage by appliance. For showering, you can estimate using 2.5 gallons of water per minute. At the end of the day, gather your

jugs and notes and determine how many gallons of water you used. This will give you an idea of your daily water use.

Note: Try not to alter your routine. This can lead to inaccurate results. If you do not have the time for this exercise, you can use the handy water calculator from the U.S. Geological Survey to get a rough estimate of your water use, available at **http://ga.water.usgs.gov/edu/sq3.html**. This calculator allows you to input your own usage by category, so you can see exactly how much water you use in a day.

Now that you have a list of all the water used in a day, consider areas in which you can reduce water consumption. If you are washing plenty of dishes — by hand or in the dishwasher — consider adding paper plates and disposable options in your emergency kit. This might bother you if you are trying to be as eco-friendly as possible, but keep in mind, they are to be used only in a true emergency, when you have no water supply in your home.

If you are worried about using large amounts of water for cooking, make sure your pantry has a selection of foods that can be cooked with little or no water. After you use the items, replace them so they will be available in an emergency. You might also want to add a "solar shower," or bag shower, which is available at camping supply stores, to your emergency plan. Used for camping, these systems store water in a bag the sun heats. When you need a shower, you empty the bag through a small hose or opening. Most are at least 5 gallons. If you use the water carefully and spend only minutes rinsing off, it can cost as little as $10. *Find sources for solar showers in Appendix A.* This is a way to ration shower water in an emergency. Also, if your water supply is cut off, your regular shower will not work, no matter how much water you have on hand.

Special needs or circumstances might require more planning and water. The following are some common considerations:

- If you have an infant who is formula-fed, determine the amount of water used each day to mix the formula.

- If you have a breast-fed infant, make sure the mother has extra drinking water.

- If anyone in your family is on any type of medication that might require drinking more water or makes them prone to dehydration, add enough water to fill those needs.

- If you live in a warm climate, you will need extra drinking water.

- If you have young children, you might need extra water for baths, extra cleanup, or in case of spills.

▸ A 5-gallon reusable jug used to store water

After you consider all situations unique to your household, make a plan. Using your notes from the earlier exercise, decide on an estimated amount of water each family member needs per day; this should include drinking water and water for personal hygiene. Each member might need the same amount, but some might need considerably more. Factor in water for pets and household use, including cooking and cleaning. Add these amounts to determine your daily water need. Then, multiply this by the number of days you want to be prepared. *Look on the Assessment Form in Appendix C for a place to record your emergency storage and use plan.*

Each household has unique needs. If it seems too complicated to make an exact calculation, you can always go back to FEMA's recommendation of storing at least three days' worth of water with a gallon per day, per person. However, if you performed water calculations, you might find your need is greater. Make sure you are comfortable with not having nearly as much water to which you are accustomed.

If you use municipal water, you can find information about your water use on your water bill, which will tell you how many gallons of water you use in a one-month period. This will be several thousand gallons or more, so it is unlikely you would be able to store an entire month's worth of water without a major restructuring of your plumbing and a multi-step rainwater harvesting and filtering process. However, you can still use the information from your bill as a baseline in your planning.

▶ An example of a household water meter

In an emergency, you will give up many water-intensive activities, such as watering the lawn, washing the car, washing multiple loads of laundry each day, and taking long, hot showers or baths. If you live in an apartment or rental unit, ask your landlord how to access the meter. There are several different types of meters, so look into how to read yours. Water meters are never reset, so you will need to do a bit of math to determine your daily use. Record the number on the meter one day, and come back the next day at the same time to find the new number. Subtract the previous day's number from the current number to determine how much water you used the past day. For example:

- On Monday at 10 p.m., your water meter reads 63,158.
- On Tuesday at 10 p.m., your water meter reads 63,558.
- Subtract Monday from Tuesday to find your daily use (63,558 − 63,158 = 400).
- Note that water meters measure in cubic feet, but they might measure in gallons. There are about 7.5 gallons per cubic foot.

Hold a family challenge to try to reduce your water use. Take timed, five-minute showers, skip doing laundry or running the dishwasher if possible, turn off the sprinkler system, and live like you only have a small supply of water to get through the day.

Then, read the meter again and see the difference between that day and a "normal" day. This limited consumption day will help you determine how much water you need to store per day, and you might determine you need to learn to conserve even more.

Planning for events that affect water supply

The first consideration of your emergency water plan should be the type and frequency of natural disasters that occur where you live. Any type of natural disaster can cause a disruption of your usual water sources, but the examples illustrated in the following sections are some of the worst for which to plan. For all these emergencies, FEMA recommends at least a three-day supply of water. Consider, too, during an emergency, you might have to leave your home, so plan for a portable supply of water for insurance.

Hurricanes

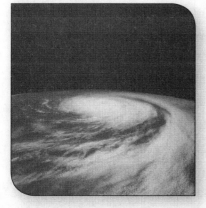 Hurricanes quickly do extensive damage to infrastructures, such as water treatment facilities, and cause widespread flooding. This flooding often contaminates local water supplies, and this contamination can persist for long periods of time. For example, after Hurricane Katrina made landfall on Aug. 29, 2005, households in most parts of New Orleans were under a boil order until Dec. 9, 2005. Katrina was a large-scale disaster,

but even a low-grade storm can cause damage to water supplies. Imagine relying on bottled water for several months if you live in a hurricane-prone area.

Consider, too, during a natural disaster, you might not have adequate transportation to water collection sites. Even then, supplies could be extremely limited, or you could encounter washed-out or flooded roads. If you rely on a well for drinking water, a major flood can quickly contaminate your well and make the water unsafe for drinking. You will have to test, and treat, the water after floodwater recedes before you can safely drink it. If you live in an area known to have large or frequent hurricanes, it would be in the best interest of your family to store as much drinkable water as possible.

If you live in a hurricane zone, you are familiar with watches and warnings issued during each season. The National Hurricane Center will issue a "Hurricane Watch" 48 hours ahead of the storm, so you have plenty of time to prepare or evacuate. Modern weather-tracking systems allow for accurate predictions of where and when hurricanes will hit land and how strong winds will be. Do not wait for the first hurricane watch of the season to start amassing water, food, and emergency supplies. Plan ahead in the months before the season, and store enough water to meet the needs of your household. If you have used any of your filled containers, be sure to top them off from the tap at the first sign of a hurricane. Also for more thorough emergency-preparedness, visit the U.S. Department of Homeland Security website at **www.ready.gov** for a complete list and instructions to assemble an emergency kit for your family. This resource will help you determine how much food, personal supplies, and stored water to get your family through emergencies.

 TIP: After the hurricane watch is set, fill your sinks and bathtubs with water. This will help meet your washing and personal hygiene needs after the storm.

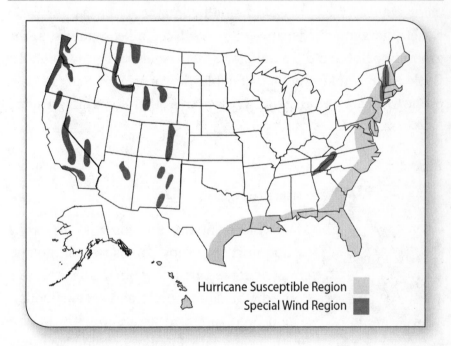

Hurricane Susceptible Region
Special Wind Region

Earthquakes

Earthquakes are another disaster that can wreak havoc on water supply systems. Even a small tremor can be disruptive to local water systems, because it could burst pipes and damage water treatment facilities. Roads and bridges could sustain major damage, which would make it impossible to get to water distribution centers for an extended time. If you frequently experience earthquakes or live near a major fault line, make sure you have the suggested three-day supply of stored drinking water.

You will also need to consider your storage options in an earthquake-prone area. Small containers stacked on top of one another could fall and break during an earthquake, which would leave you without water when you need it the most. Consider using a variety of large and small containers you can store in closets, garages, sheds, and outdoors. Keep all containers on the ground or floor, and do not stack them unless it is unavoidable. Look for durable, thick-walled plastic containers that are less prone to breakage if they fall over or have other items fall on them.

Floods

▶ Flooded playground

Many places across the country are prone to flooding, including towns near large and small rivers and areas that hurricanes affected. Some dry areas also experience "flash floods," which occur after an area suddenly floods, due to heavy rain. Flooding could also create a need for "extra" water. When rivers, lakes, or waterways overflow their banks, they often pick up debris, bacteria, and pollutants, such as gas or oil, along the way. This water can then flow into wells and city water systems and make the water unsafe for drinking and many other household applications. After this water is compromised with floodwater, it could take weeks to return the water to a potable state. If your area is prone to flooding, store your water in sturdy, non-porous plastic containers with watertight seals. Look for containers that will keep floodwaters out if they are submerged. If your water containers are submerged in floodwaters, even if the seal is not broken, you will still want to treat the water before you drink it for safety's sake. *See later sections in Chapter 1 for instructions on how to safely purify water through boiling or a chlorine bleach treatment. Chapter 6 provides instructions for more advanced treatment methods.*

Droughts or water shortages

The opposite end of the spectrum — but just as serious — are areas that suffer from droughts or water shortages. If your grass dies, you will not be in a life-threatening situation, but you can use gray water to keep at least a few plants alive if you are under a watering ban or extended drought. *Find complete information on reclaiming gray water in Chapter 4.* To use your gray water for plants, capture the water when you bathe or shower by stopping the drain. Run a regular garden hose out of the bathroom window to a waiting container or to the plants you want to water. Then, siphon the water out, and use it on your non-edible plants. Never siphon the water by mouth. Start the siphon by running water from the faucet to fill the hose. Hold a section of hose above the faucet. While the water is still flowing, quickly take the end from the faucet and submerge it in the bath water. Another way to siphon is to fill the hose with water and block the end that is outside. Submerge the other end in the bath water. Release the block on the end outside and the water will create a vacuum and siphon out the bath water. Do not store this water; only use it to water plants and grass.

Tornadoes

Tornadoes occur almost everywhere in the United States and, like hurricanes, heavy rainfall and high winds that cause damage to infrastructures and roadways accompany them. Unlike hurricanes, tornados crop up quickly and leave the homeowner just enough time to seek shelter and no time to gather emergency supplies. These massive storms can also disrupt local water supplies, which makes it essential to have enough water on hand to weather the days after a storm. Make sure to store at least some of your water in the same area in which you will seek shelter in case of a tornado. Then, if you are trapped in this space, you will have access to water until help arrives. Remaining water can be stored in a different location; just make sure it is in an underground area that will not sustain damage from a tornado.

Ice storms and blizzards

Winter brings with it another weather risk in the form of blizzards and below-freezing temperatures. Heavy snows, thick ice, and high winds can cause power outages that will affect your ability to pump water if you rely on a well. Extreme and prolonged low temperatures can burst pipes and cause your home to be without access to water. Repairs to these broken pipes are hampered further by having to work through frozen ground and chilling outside temperatures. In addition, roads could be impassable after a heavy snowfall, and you will not be able to get to a store to buy water. Again, storing a three-day supply of water will get you through even the harshest snowstorm. Water stored for winter use must be kept in an area that does not freeze. Otherwise, you will find blocks of ice or even worse, ruptured containers from expanding ice blocks.

Map of average snowfall in the United States

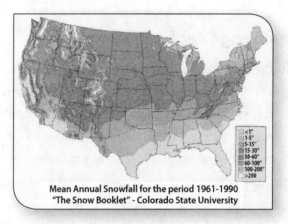

Mean Annual Snowfall for the period 1961-1990
"The Snow Booklet" - Colorado State University

▲ "Water and Climate Program, Products and Services Briefing Book, National Snow Data Management System Initiative" from the Natural Resources Conservation Service (NRCS), a division of the U.S. Department of Agriculture, at **www.wcc.nrcs.usda.gov/publications/Briefing-Book/bb27.html**.

Manmade disasters

Even if you live in an area that does not have natural disasters, store water in manmade emergencies. Stories from throughout the country tell of contaminants accidentally entering the water systems and making tap water undrinkable. Aging infrastructures and water mains break and leave city blocks without water for days. Unexpected high winds or falling trees can create power outages and make pumping water impossible. Also, there is the threat of terrorist activity that could render the water supply unsafe. All these contingencies are addressed in most city plans, but it might take days or weeks to repair damage to water systems, leaving the typical homeowner without access to usable water. Planning ahead with a few days' supply of potable water will save you from last-minute disasters.

Disaster storage techniques

When planning for emergencies, consider your household water system. If you rely on a well for your household water, know what is involved in testing the well water for safety reasons. Contact your local health department for information about testing your well water. If you have city water, consider the reliability of the water supply. Having your own potable, or safe to drink, water available will make you more self-sufficient and give your family a valuable tool for getting through difficult situations. Safe, available water can even mean the difference between life and death in a serious natural disaster. *Following chapters will cover various methods of testing, purifying, and using water during an emergency.*

The main concern with storing water for emergencies is protecting the purity of the water while it is stored. Unless you plan to have a purification system in place, you will want to make sure your drinking water is kept in safe and clean containers. Even if you have rain barrels, a pond, aquifers, and a swimming pool, you will still want to have at least a three-day supply of potable water in sealed jugs or barrels. In the midst of a disaster, you do not want to have to put a complicated purification system to work — you just want to have water readily available.

Smaller water containers, anything from 2-liter soda bottles to 5-gallon water containers, can be stored in a variety of places. Put them on the floor in closets, in the garage, sheltered on a porch, or in a shed — any place out of the way. You can even stash cases of bottled drinking water from the store under beds, behind the couch, or anywhere they will fit. If you have a safe home water supply, you can fill up your water containers from the tap.

If your water is treated with chlorine, you can close up the container. If you take water from your well, add two drops of liquid chlorine to the water to sanitize it, close it up, and put it in your chosen storage location. Also, write the date on the container to rotate through your stored water. It is recommended, but not required, to use your water within six months of storage. Be aware, though, the safest water to store is water bottled in a factory and left sealed in your storage area. When you fill your own containers, you are likely to introduce bacteria or not properly sanitize the container.

A final consideration for those in hurricane and flood-prone areas: If your water containers are easily portable, you could take some with you if you are required to evacuate your home. If you evacuate to a shelter in the immediate area of the disaster, the additional water could make the situation more comfortable for your family until you can return to your home.

No place is completely free of emergencies and disasters. Even if the chances of natural disasters are small, the possibility of manmade disasters — including riots, strikes, and industrial accidents — are still there. By making a solid emergency water plan, you can ensure less stress in an emergency in your household. You can be confident whatever happens, your family will have an adequate supply of safe drinking water in any situation.

Helping others in a time of crisis

So far, the benefits of water storage have been discussed in relation to how it can help you. However, storing water can help those around you in a time of crisis. If your community and neighborhood find themselves in an emergency situation, you might have more than enough water stored for yourself, and you can then offer excess to others.

Communities should help one another out in emergencies, and providing water is an excellent way to assist those around you. If a neighbor's well pump stops working unexpectedly, your neighborhood loses electrical power, or winter temperatures freeze or burst pipes, it might be many days before people in your community have access to water. Even though the emergency might not affect you because of your advanced preparations, it is comforting to know you can help someone in a time of need. You never know when you might find yourself in a similar situation.

Another thing to consider is, you might not need to work individually on your water storage preparations. A number of communities around the United States work together to build water storage units. If you have a limited amount of space on your property and your neighbor has a large lot, consider building a cistern to benefit both households. By working together, you share the cost and create a system everyone can use.

CASE STUDY: WATER STORAGE IN THE DESERT

Carole Crews
Gourmet Adobe
www.carolecrews.com

I have a cistern at my dome house in the desert, and all the water from the roof goes into this cistern for household use. It is pumped into the house with a 12-volt pump run by solar power. My neighbor drinks her rainwater after running it through an extra filter or two. She has a 5,000-gallon cement cistern built right into the ground, and it has worked extremely well. I also have a pond on my main property that catches water from the roof and irrigation water from snowmelt in the mountains. The pond at my other property is used to keep gardens going between irrigation day, and I also send gray water to garden areas from several drains in both houses.

This stored water is my only source of fresh water here in the desert, as we have no possibility of drilling a well. I also believe these water systems saved my life when a forest fire was threatening to come close to my house. I used the water to wet the area down, and we avoided that catastrophe.

One of the benefits of using stored water is it is often superior water for gardening and washing, because it has no chlorine or other additives. I also believe the less water we pull from beneath the ground, the more ample our water tables will be for future use. Some of the other benefits are a sense of security in knowing one has enough water, being able to grow food with it, and being less dependent on large systems that cost money and go haywire.

I did, however, need some particular materials to set up my water systems. For my pond, I only needed a pond liner, a backhoe to dig the hole, some pipe for a downspout, and gutters. At the dome property, I needed to purchase a large cistern, bury it, set up pipes to carry the water, and use roofing materials to keep the water clean. The solar electric system runs the small pump.

People should store water at home for independence from large systems, security, and lushness in their gardens. One of the most important things in emergency water storage is to make sure to have clean containers that can be tightly closed if you plan to drink it. For bathing, I save water by keeping it in a large, wood-fired, stock-tank hot tub and clean it up with soap before I enter the hot water.

Storing Water for Household Use

There are numerous ways to use rainwater or gray water for household use — especially for landscape watering or non-drinking purposes. *These methods will be discussed at length in following chapters.* Capturing water from these sources can save money, conserve water, or create a more "green" lifestyle. Others store water for household use because they do not have ready access to water because of where they live, such as in remote, rural areas. These people rely on wells, cisterns, and other water collection and storage systems to provide water for their families, livestock, and vegetation. Any water that does not come from a municipal source must be treated or purified before you can drink it. Choosing to store rainwater or other natural water also reduces your reliance on local water systems, which will significantly reduce your water bill. There are a wide variety of water storage containers available, and unless you have specific needs — for earthquake or flood resistance — you can choose whatever will work best for your household. *See following chapters for collection and storage options.*

Using a well to access water from an aquifer

If your household has access to a well, you are collecting water from deep underground — from the area known as the aquifer. An aquifer is a layer of underground, porous rock, gravel, sand, or clay, which filters groundwater and transmits it below the groundwater level. The water in an aquifer can run free and recover by a well. These aquifers can eventually recharge if water is able to reenter the underwater basin. This means water enters at a higher elevation on the basin through porous rock, coarse sand, gravel, or openings in the rock. The aquifers are often sloped, and therefore, the water runs in a downward flow and filters into a sort of underground river. To fully recharge an aquifer, though, the water must refill at a rate equal to or greater than the discharge flow. Furthermore, new water entering an aquifer is not as well filtered or "pure" as the water held in the aquifer for hundreds or thousands of years.

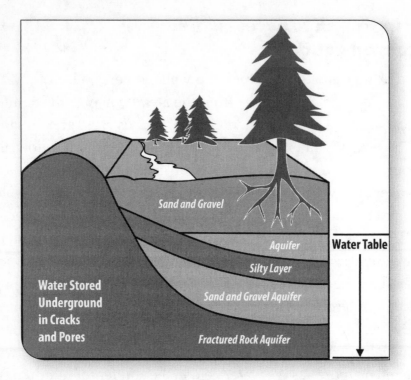

There are many identified and well-recognized underground aquifers in the United States. In some cases, an aquifer is the size of an underground ocean, such as the Ogallala Aquifer that runs from Nebraska to western Texas. This aquifer contains water from 10,000 years ago and is a predictor of the dire consequences of depleted water supplies.

The Ogallala, also known as the High Plains Aquifer, sits in the heart of America's agricultural land. It is estimated that 90 percent of the water removed from this aquifer is used to water one-fifth of all U.S. cropland. Additionally, because the rainwater must flow over farmland to go back into the aquifer, much of runoff is contaminated with eroded soil, manure, and leached farming chemicals, such as nitrates and pesticides. Although numerous experts keep watch over this aquifer, it is not entirely known how long the water supply will last or how much time it takes to completely recharge it.

Reaching the proper depth in an aquifer when drilling is important and varies by region, because the water table is not stationary and can move up and down over time and in different conditions. In some parts of the country, a shallow well might dry up if there is a drought. Deeper wells can access water further down and are more reliable. The water in a deep well, too, has been there longer and has better quality than that reached in a shallow well.

The type of well providing water to your home and property depends on the particular area in which you live because the water tables and depths of aquifers differ from one region of the world to another. In some regions of the world, it is easier to drill all the way to an aquifer, and in others, it might be too deep or embedded in deep rock. Wells that reach into the aquifer are often referred to as artesian wells. To qualify as an artesian well, the pressure of the water must be sufficient to push the water up the pipe above ground level. These types of artesian wells are considered free-flowing, which means water will flow upward toward the surface without the need for a pump. Impermeable rock does not confine other types of aquifers, and therefore, they do not have sufficient pressure to push the water up the pipe to the surface of the well. These type of wells require a pump in order for the water to come to the surface.

▸ Inside of a gravity well

Gravity wells, referred to as shallow wells, only reach as far as the water table. Gravity wells require water to be pumped to the surface, either by a hand pump or a mechanical pump, such as a windmill. If your power fails and you have a well that requires an electrically run mechanical pump, you might find yourself without water unless you have water stored for emergency use.

Few homeowners, however, have the equipment or expertise to dig a new well or refurbish an abandoned well. These jobs are best left to the experts to ensure your well will provide safe water. First, you will have to get a permit for your well, so check with your local zoning office for information on the steps required. To find a contractor to dig your well and get it running, ask your local hardware store or area builders for recommendations on whom to hire for this job. Any certified well digger can provide you with a well, but it is best to hire someone familiar with your area. He or she will know the ground conditions and can help you find the best location and avoid surprises when the excavator arrives. Throughout the process, an inspector will need to visit the site — check your permit for more information. Additionally, after the well is dug, test the water quality before hooking it up to your system. New wells often require periodic testing and maintenance. Your contractor and inspector will let you know the steps required to keep your water safe and your well running smoothly.

Supplemental backup storage

Having a deep well is important, but you cannot always control variables, such as earthquakes, construction, water formations below your property, the depth of your water table, and other factors that can change the course of groundwater. Planning ahead and having alternative sources of water can be a lifesaver in an emergency. If you have a well, a supplemental storage tank for your water is still an idea. If your well pumps water into a large holding tank before it is used in the house, you will have an emergency water supply in case of a power outage. Unless your well pumps water via a windmill or solar power or you have a gas-powered generator, you will have no water if the electricity goes out. Depending on the size of the holding tank, it could provide several days' to a week's worth of water for your family.

Even if you have an alternative water system, such as a rainwater harvesting system or a gray water reuse system, consider being connected to the local water supply. If there is a prolonged drought, your rainwater system could run out of water. In a gray water reuse system, you still need fresh water for drinking and also for the first-time use in showers and washing machines. A dual-access water plan, with alternative and traditional water supplies working together, is a way to ensure you will always have access to fresh drinking water, no matter what situations crop up. This book will walk you through your many options in storage, gray water use, and rainwater collection. Your imagination, effort, and budget only limit the combinations of these alternatives.

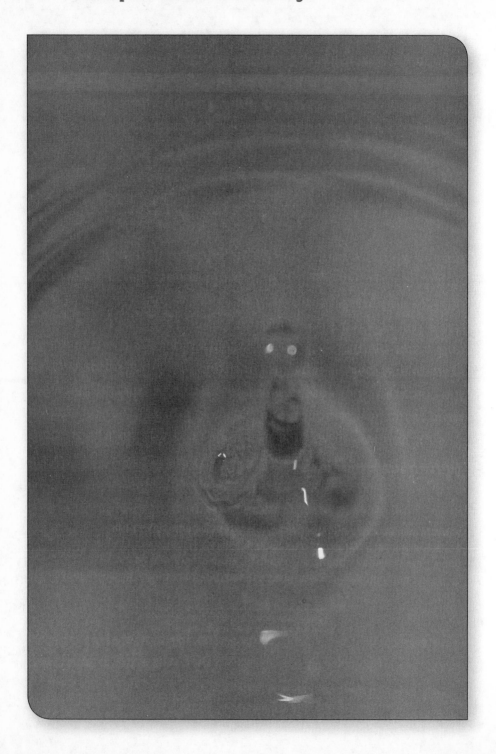

Common Sources of Water

If you live in an area a municipal water supply serves, the fastest, easiest, most reliable source of water you have is right at your tap. This water goes through the highest level of testing, it is conveniently routed to your home and all your spigots, and you pay a relatively low fee for this precious commodity. Of course, there are many times when this source of water is compromised or unavailable. *As discussed in the previous chapter, municipal water supplies could be contaminated, or pumping stations could be damaged during disasters, and you will no longer have ready access to clean, running water.*

? The Issue of Fluoride

The Centers for Disease Control and Prevention (CDC) estimates that nearly 60 percent of homes in the United States receive fluoridated water. This fluoride is naturally derived from fluorine in the water source and is added as part of the water treatment process. Fluoride is proven to reduce tooth decay, and widespread municipal water fluoridation is credited with significantly reducing dental diseases. Fluoride is considered especially important for children as their teeth are formed. There are, however, opponents who question the safety of adding fluoride to water, but prominent health organizations, including the American Dental Association (ADA) and the World Health Organization (WHO) still back this practice. If you collect your water privately and run it through a purifier, you might remove this essential element. It is possible to get too much fluoride, though, so do not attempt to add fluoride to your water unless you have checked with an expert or your family's dentist. To read up on this topic, visit **www.kidshealth.org,** and search for "fluoride."

Additionally, municipal treatment centers rely on power and chemicals to keep this water clean, and homeowners use and waste this water for non-drinking purposes, such as watering the lawn or washing the car. This highly treated water is too clean for this sort of use, and increased populations are beginning to overtax aging treatment facilities already at their peak use. If you depend on a well for your water, this same theory applies and is even more important to consider when you use your water. Any water pumped from the water table and used to water grass is not water utilizing its full potential. Every additional demand put on the already strapped groundwater supply makes this supply even more diminished and difficult to access. These reasons are why the best sources of water for most household uses do not come out of your tap, but from rainwater and "used" water known as gray water. *Remaining chapters of this book will delve into the best ways to harvest, purify, store, and use the sources of water.*

Rainwater Harvesting

▶ Rainwater barrel

The simplest home water storage involves rainwater harvesting. As its name suggests, rainwater harvesting is the collection, storage, and use of rain. Areas prone to drought have used rainwater harvesting for hundreds of years, and this water collection method has gained popularity across the nation. The most obvious benefit of rainwater harvesting is that it is nearly free. There might be some initial cost in setting up the equipment required to collect the water, but the supply itself is cost-free and readily available in most regions. Fortunately, as the popularity of rain barrels increases, associated costs are decreasing. A few years ago, you had to special order rain barrels, but now you can find them in many large garden and home improvement stores. Although you can buy rain barrels online, due to the size and weight of a rain barrel, buying them locally can save you money associated with shipping costs. In addition, many municipalities, organic cooperatives, and community organizations are now selling rain barrels through special sales. Check with your city for more information.

Another benefit of harvesting rainwater is that it is naturally distilled and a pure source of water until it comes into contact with contaminates. Distilling water is a process that removes contaminants or impurities by boiling the water. The water condenses into steam and leaves the impurities behind. When water droplets form, the collected water is purified. With rainwater, this process happens naturally. Clouds distill the water before it rains, and the resulting water contains few dissolved solids, such as calcium and other minerals, compared to water from surface sources.

Because the water contains fewer minerals and solids, it is considered soft water. Soft water contains less magnesium and calcium compared to hard water. These elements can interfere with the formation of lather in soap. This means with soft water, you use less soap when washing, which in turn reduces the amount and chemical concentration of gray water your home produces. It also reduces the length of time it takes to shower or clean clothes. Rainwater is also considered a natural hair conditioner. Some health and beauty companies sell shampoos and conditioners that contain rainwater.

In addition to rainwater being soft, it also contains the least amount of salt of any other freshwater source. Many salts occur naturally on the Earth's surface, including sodium, potassium, magnesium, and calcium salts. Soil with high concentrations of these salts can inhibit the growth of plants because it reduces their ability to absorb water. Rainwater can dilute these salts, whereas surface and groundwater spread salts in soil. You can use harvested rainwater to flush the salts out from around plant roots to help them grow and absorb water more efficiently. Rainwater can be used as a natural fertilizer because it often contains sulfur, a beneficial element for healthy growth. In addition to sulfur, it contains beneficial microorganisms and other natural minerals that help plants grow greener and healthier.

How much water can you collect with a basic system?

To start a system to collect rainwater, redirect your gutters into a holding tank. You can harvest more water than you might think. Only 1 inch of rain falling on a 1,000-square-foot roof will add up to more than 600 gallons of water if it is all caught. However, it is not possible to harvest all the water because some will be lost to splashing or evaporation as the rain hits your roof or travels through your gutters. The following is a simple formula to calculate your potential rainwater harvest.

> Catchment area of building in square feet × Inches of rain ÷ 12 × 7.48 (to convert to gallons) = Actual amount of rain falling on roof. Assume at least 25 percent of this sum will be lost to evaporation.

An example is 1,000 square feet with 1 inch of water:

$$1000 \times .083 \times 7.48 = 620.84 \text{ gallons}$$

Taking away 25 percent yields about 466 gallons of water from one rainfall. This is a substantial amount of water that would require a large storage system in place during each rainfall. Most home systems available for sale at home improvement stores are made up of a 55-gallon barrel or two, which catches enough water for watering houseplants or container gardens. This rainwater harvesting system collects the water running through the roof gutter system and empties it into the barrel. A spigot near the bottom of the barrel releases water into a watering can or bucket for use on plants. More complex and advanced systems use larger tanks or cisterns and involve filtering, purification, and pumps to get the water out of the tanks. Depending on the needs of your household, your annual rainfall, and your interest in rainwater harvesting, you have many different options from which to choose. *Chapter 5 discusses rainwater harvesting techniques in more detail. Appendix D also includes project plans for creating your own system.*

Precipitation (in inches) across the United States in 2009

▶ "AHPS Precipitation Analysis" National Oceanic and Atmospheric Administration (NOAA), a division of the U.S. Department of Commerce, at **http://water.weather.gov/precip**.

This map shows how many inches of water fell during 2009 in areas across the United States. Use this map to determine about how many inches of rain you can expect to harvest in your particular part of the country. In drier areas, water storage and conservation are even more crucial.

CASE STUDY: I BUILT A BETTER RAIN BARREL

Barry Chenkin
Aquabarrel LLC
554 N. Frederick Ave. #122
Gaithersburg, MD 20877
info@aquabarrel.com
www.aquabarrel.com
301-253-8855

While watering the plants with the garden hose one day, my wife asked if I knew what a rain barrel was. To my dismay, I had to admit I had no idea what she was talking about. She explained the concept. With my interest peaked, I spent many hours researching rain barrels on the Internet. Finally, I decided it was possible to make my own version. Following the traditional concepts of a rain barrel, I proudly made my own.

When the first thunderstorm came, I was giddy with excitement because my creation began to fill up with precious water. I had never paid attention before this to just how much water could accumulate during a rainstorm. My excitement soon turned to shock as the rain quickly filled the barrel and began to flow over the top, run down the sides of the barrel, and pool against the foundation of our home.

Watching this soggy disaster unfold convinced me a few significant modifications were in order. Working with my original rain barrel taught me the flaws in the traditional design. So I tweaked this design with an assortment of modifications. I also decided I was not the only one wanting an easy-to-use rain barrel, so I started my own business selling this and many more rain collection systems. Who knew watering the garden would lead me to a business venture?

During this process, what was the biggest lesson I learned? Choose a rain barrel that can handle the amount of rain that will come into it. You will not capture all the precipitation, so know where the excess will go — or make sure you have an overflow port, because it's not going to stop raining once your rain barrel is full. Next important lesson as a rainwater harvester? You must periodically drain the barrel of the bottom few inches of residual water or built-up sludge or plan on opening a mosquito harvesting business.

We were also glad we had access to collected rainwater when storms knocked out the power to our well pump. That was a lifesaver for flushing toilets, bathing, and keeping things going until the power came back on. When we have power, I use this retained water for non-potable uses, such as washing vehicles, irrigation, and filling a small pool for my grandson to play in. One of the biggest advantages to collecting rainwater is it reduces the amount and velocity of storm water flowing off your property. This, in turn, alleviates downstream flooding, erosion of stream banks, high levels of pollution, and trash carried into sewers. This is especially beneficial in urban areas, where most storm water ends up in waste water treatment plants.

The only thing I'd like to change about our system is to install a different roofing material. We have those nasty asphalt shingles, and the messy residue is picked up and carried into the rain barrel. For smaller tanks or those without a first-flush filter, this creates a problem in the bottom of the barrel and for the water quality. We are looking at replacing our primary collecting roof with steel.

Purification of rainwater

Setting up a rainwater harvesting and storage system is easy, but if you want to be able to drink the water, you need to purify it. If you do not use proper purification equipment and procedures, the bacteria and contaminants in the rainwater could cause serious and even life-threatening illness. According to the EPA, boiling water is the best way to kill bacteria. Boil vigorously for one minute, or three minutes at altitudes more than 5,280 feet, to make the water safe for drinking.

 Boiling water for drinking, personal hygiene, and household cleaning is a short-term solution to purification, but it can be time consuming and labor intensive. In addition, boiling water relies on power sources that might be inaccessible during emergency situations when you need water the most. Although wood stoves and camping stoves will work without electricity or gas, your fuel supply is still limited, and keeping quantities of fuel on hand can be expensive. Because you will want to rely on the same system for household use, you need to plan with that in mind.

An easier, but less effective, method is to treat the water with household chlorine bleach. According to FEMA, adding 1/8 teaspoon (or 16 drops) of bleach for each gallon of water will kill most of the disease-causing microorganisms in the water. Use only unscented, plain household bleach, and use a new or unopened bottle if possible because bleach becomes less effective with time. Add the bleach to the water, stir, and allow it to sit for at least 30 minutes before use. When handling bleach, be sure to work in a well-ventilated or outside area as the fumes can quickly injure your respiratory system. Wear long rubber or plastic gloves and goggles to protect from splashes — contact with bleach can cause injury. Store your containers of bleach in a cool, dark place away from children.

Boiling and chlorination will kill most microorganisms, but it will not remove any debris or heavy metals in the water. Rainwater running off a roof can pick up tiny particles of debris from the shingles or other roof surface. For either situation, make sure to filter the water through some clean cloths first to remove any larger particles or debris. For example, pouring the water through a clean T-shirt will remove the debris.

These methods are acceptable for an emergency situation, but if you want to have filtered rainwater available for everyday use, you will want a filtration system that requires minimal time and the best filtration levels possible for the health of your family. *Chapter 6 will discuss purification and filtration in more detail.*

Gray Water Collection

Another type of water that can be captured and stored for household uses is gray water. This type of water comes from baths and showers, washing machines, bathroom sinks, and sometimes, if you do not have a garbage disposal, the kitchen sink. Water from the toilet, called black water, can never be reused in your household.

Gray water is channeled from the pipes in your home to an external pipe or holding tank and used to water trees, bushes, and plants. Gray water cannot be used for drinking or on edible plants without extensive purification. Some gray water designs also create a drain field, where the gray water will flow onto the land and naturally filter back into the groundwater. The main benefit to using gray water is the water is used to its maximum capabilities. Most water the typical household uses daily is fairly clean after its first use but still runs down the drain to a municipal sewer system or private septic tank. Even the dirtiest dishwasher load creates water that could be used again to water plants.

Without gray water collection, this water runs back into the city treatment plant and is inundated with chemicals and processing. By using gray water, you are getting two uses out of the same water and reducing the strain on these septic system and water treatment plants because less water is going into it. Removing gray water from the treatment cycle helps to reduce the amount of water wasted. If your home is connected to a city sewer system, capturing gray water for household use will reduce the amount of energy used to pump waste water to a treatment plant and process the water. Furthermore, if you are charged a "sewer" fee by the gallon, you will save money because less water is leaving your home.

A gray water system can also offset the needs of a private septic system. By setting up a gray water system to redirect the gray waste water away from septic storage, a smaller, simpler septic system would be needed. This is especially important for situations where installing a large septic tank is not feasible, such as with limited space or slow soil percolation, which is the soil's ability to absorb water.

The upper layers of soil, where the microorganisms break down toxic materials through the biological process called bioremediation, will filter and purify gray water spread on the plants and lawn. According to the EPA, bacteria in soil consume toxic materials, such as metals, oils, and even gasoline. The bacteria transform the materials into water and carbon dioxide. This is what protects and filters surface and groundwater, so using gray water for irrigation purposes purifies it more efficiently than standard water treatment plants. This natural process also quickly replenishes any nearby aquifers, keeps groundwater levels up, and reduces the impact of drought conditions. Additionally, people using these systems output "cleaner" gray water because they are less likely to dump contaminants, such as paint, household chemicals, or automobile fluids into the waste water stream they plan to reuse.

How can I use gray water?

Untreated or unpurified gray water should not be used for drinking purposes, but it can be a useful part of your household water supply. Gray water is either sent to a holding area outdoors or it is set up to drain directly onto certain trees or areas of the yard. You can use and direct the gray water in a number of ways, including creating a water garden in a holding tank or directly diverting the water from your appliances onto the soil.

TIP! Using biodegradable, phosphate-free soaps and natural products in the shower and laundry can help keep gray water a safe source of additional water for your landscape. Think of it as another form of recycling — one that will keep your yard healthy and well watered.

 Irrigation is the most viable use for gray water. Soaps and chemicals used in the home put phosphorous and nitrogen into bath and washing machine water. Many plants need soil rich in these ingredients, so gray water can benefit them and be applied directly to the areas that need frequent watering. Many sources agree gray water is fine for irrigating ornamental, non-edible plants, such as lawns and decorative gardens. Experts are split on the use of gray water with vegetables and fruits, and it is always better to err on the side of caution when it comes to the food you eat. In an ideal system, gray water sources are sent to one portion of a septic tank — black water has a separate section — filtered through sand or soil, and made available for irrigation purposes.

Another way to naturally filter gray water is to direct it to a pond or holding tank filled with live vegetation. With these systems, water plants form a type of mini-wetland area, and these plants, in turn, can help clean the water before it enters the filtration system. When combined with the use of biodegradable soaps, water gardens give you high-quality water to use for your plants and yard. Large livestock tanks, smaller pre-formed pond liners, and even old bathtubs are frequently used to create the water garden or wetlands area. These holding tanks also have the added advantages of creating a green space and catching additional water when it rains. The addition of a water pump makes it easy to remove water from these storage areas, but many people use a bucket to take water out or set up a system of overflow hoses to direct water to areas that need it the most.

Gray water is not immediately usable for ponds with fish due to the possible presence of soap or other household cleaning agents. With proper filtration, it is possible to use gray water for your aquatic wildlife, but it takes some specific knowledge. Check with your pet supplier or a marine vet for more information. Gray water can be used in a pond filled with only water lilies or other vegetation, but this depends on many factors, including: the size of your pond, how frequently gray water enters the pond, if you are using aeration, and what type of vegetation you are planting. Frequent influx of soapy gray water into a small barrel might be too much for plants to handle. Consult with your garden center to ensure you make the right selections for use with gray water systems. For more information on pond building, refer to the book, *The Complete Guide to Building Backyard Ponds, Fountains, and Waterfalls for Homeowners: Everything You Need to Know Explained Simply*, also a product of Atlantic Publishing. Additionally, before you decide on a pond system, check your local laws and ordinances to make sure your particular system is legal.

There are also smaller, stand-alone gray water systems to use in the bathroom or laundry room — one example is a small system that mounts under the bathroom sink and captures water from the sink and tub to flush the toilet. The complexity and cost of these systems range widely. Fully automatic gray water recovery systems cost about $200 and require the same plumbing skills as it takes to install a new toilet. A simpler manual system, such as using a bucket to collect shower water and then pouring it into the toilet holding tank, costs next to nothing but takes more time to manage. You cannot use water from the toilet itself, but you can use gray water to flush the toilet without any health concerns.

Using a bucket to capture some of your household water is the most basic way to use gray water. A bucket in the shower while you wait for the water to warm up can catch a few gallons a day you can use for flushing toilets or watering plants. You can also put a bucket in the kitchen sink to capture water from washing your hands or running water while waiting for it to heat up. Although catching water with buckets will not save as much as an entire household gray water system, it should provide enough water for your plants.

Legal and Permitting Issues

Systems to collect water have always been in place for those living in drought-prone areas, and over the years, a variety of designs have sprung up to address specific needs of the homeowner. In addition, increasing populations in water-stressed areas have made conservation even more important for overtaxed municipalities and their water treatment facilities. Large rainwater collection and gray water harvesting are fairly new practices to most cities and counties, and their governing boards are playing catch-up with regulating these systems.

These rules are not just a way for the government to get their hands on more permit money. If done improperly, harvesting and using rainwater or gray water can have an adverse impact on cities and their residents. For example, gray water overflow or misdirection could contaminate municipal water supplies, and over-collection of rainwater could create a water shortage for neighboring properties. Because everyone's water comes from the same source, municipalities are realizing they must step in to ensure these practices are done correctly and will not affect surrounding areas or water supplies. Laws are written to help out the consumer and put restrictions on where, when, and how you might harvest water. Conversely, some areas have no rules in place. As a homeowner, the best plan is to check with your local governing bodies before you begin any project, and make sure you have all the permits in place before starting construction.

Once you have the space, the money, and the expertise to go forward with harvesting or collecting water, you must still determine if these practices are allowed by law in your area. As of 2011, no overriding federal law exists covering these issues — but that might also change as this issue moves to the forefront. Each rule now is enforced at the city or county level and is written with specific local issues in mind, such as runoff into existing sewer systems, water availability issues between neighbors, or concerns over pollution entering existing water resources. Because the issues are so different, you must find out any local rules and regulations that will affect your specific plan before you begin building any water storage or collection system.

Most municipalities are in favor of reducing the pressures on their waste water treatment plants and encourage water conservation as much as possible. The problems arise when well-meaning conservationists use harvesting and usage techniques that eventually cause damage to existing water systems or established filtration boundaries. One such example would be a homeowner deciding to use gray water for watering his lawn. He or she takes his or her untreated gray water, potentially filled with soap, chemicals, or even disease organisms from bathing. This water is applied (or over-applied) to the lawn, it runs off into a nearby pond or gutter system, and then it contaminates the surrounding area that might be feeding into a city's water supply. The same situation can occur with rainwater harvesting if the rain is collected off a particularly dirty roof or contaminated hard surface. Some states, such as Colorado, have strict regulations and limitations on the rainwater harvesting systems you can build.

?

Top Five Tips for Working With Regulators

Sarah Lawson, of Rainwater Management Solutions and contributing author of the Virginia Rainwater Harvesting Manual, recommends these tips for working with regulators to get project approval:

1) **Talk early and talk often:** No one wants to get to a final inspection and have a project turned down. Rainwater harvesting and gray water systems are still uncommon in many areas. Start talking to the people who will need to approve the project while you are in the early stages. Planning and purchasing a system that will get approval is easier and more cost-effective than buying everything and finding out you cannot get approval.

2) **Be prepared to provide education:** Because a the technology is fairly new, you might want to provide information up front about how your system will work and what steps you are taking to ensure you have a safe system.

3) **Choose your battles:** When trying to get approval, keeping a big picture view is important. Getting caught up or upset with any required design changes or restrictions can make you lose sight of the big goal, which is getting your water system in place. The last thing you want to do is get into a contentious relationship with the people approving your system over a relatively small issue.

4) **This does not need to be adversarial:** When working with code officials and inspectors, it is their job to be cautious. They have been charged with protecting public health and safety, and in the end, you want the same thing: a functional, safe system. The best way to get there is to work with the reviewers and inspectors instead of seeing the relationship as adversarial.

5) **Help the code official find allies:** Code officials across the country are trying to decide what can and should be approved, and most are reluctant to step out on a limb alone. Referring these people to other officials or cities that have approved similar systems can help them make an informed decision and feel more comfortable approving your plan. A company that specializes in water storage and collection could help point you to experienced code officials and regulators. Ask your tank supplier or contractor for help, or search online for "water storage," "water reclamation," or "water collection" specialists in your area.

One place to start for guidance is at your local cooperative extension office. These are the people most familiar with the laws and regulations concerning water storage, water catchment, and the use of gray water or rainwater systems. The USDA website, **www.csrees.usda.gov/Extension**, is a resource for finding your local office. In addition, every state is divided into Soil and Water Conservation Districts (SWCDs). These government bodies are set up to work with farmers and landowners to protect and conserve the water and soil resources through proper construction techniques and management practices. These offices will provide free or low-cost design services and construction assistance based on your individual needs. You could even qualify for federal or state cost-share in building your system, and they will assist you in securing this financing. *More information on incentives is listed in the next section.*

If collecting and storing water is not against any laws or regulations in your area, you will also need to check on building codes to make sure your storage tank and the concrete base, or footing, you place under the tank meets local requirements. The following five steps will assist you in ensuring you have checked all relevant laws and regulations:

Step 1: Start low. Check with your homeowner's association or historical district to determine whether there are rules that apply for your home.

Step 2: If your neighborhood does not have rules against outdoor water storage and rainwater tanks, check with your town or city. Find town planning information either online or at your local town or city hall.

Step 3: Check state laws to ensure you can proceed with your plan.

Step 4: After town or city approval, check building codes to determine the type of foundation required for a large water storage tank. If your town does not have specific requirements for water tanks, consider consulting with a town planner to discuss what types of foundation will work best in the area.

Step 5: Make sure any required safety features are added to your building plans.

While checking about state local regulations, you might also want to inquire about county and state level rebates and tax credits for rainwater harvesting if that is the type of system you are installing.

Cost Sharing and Tax Incentives

Many states offer tax rebates, credits, cost-sharing, or deductions for costs incurred in the installation and use of alternative water harvesting methods. The following are some examples of the incentive programs in the United States:

- **Forest Stewardship Program:** Run through the U.S. Department of Agriculture, this program offers a matching program to reimburse landowners around the country for costs incurred when building water harvesting earthworks and erosion control systems. The program varies by state, so check with your local office for more information.

- **Arizona water conservation systems:** Arizona residents can get a tax credit of up to 25 percent of the cost of installing a rainwater or gray water harvesting system. A limited amount of money is set aside for this program each year though, so apply early. Builders in Arizona can get a tax credit of $200 for each qualifying rainwater or gray water harvesting system they install commercially. To find more information about this project, including necessary forms, visit Arizona's Department of Revenue website at **www.azdor.gov/TaxCredits/WaterConservationSystems.aspx**.

- **Santa Fe County, New Mexico:** In 2003, the county passed an ordinance that requires all new construction to include some type of rainwater harvesting system. The requirements vary based on the size and type of the construction.

- **Clean River Rewards:** The city of Portland, Oregon will reimburse homeowners up to 100 percent of the cost of storm water management if they install storm water management systems on their property. This means if you can build earthworks and harvest rainwater, which keeps rainwater from running off your property and into storm sewers, you will receive a credit on your water bill. Find more information at the program's website at **www.cleanriverrewards.com**.

- **Downspout Disconnect:** Also in Portland, this program disconnects downspouts from sewer lines. A city worker will come to your house and change the direction of your downspouts from going into storm sewers onto your property for irrigation. This is done at no cost to the homeowner. Those who decide to do it themselves will get a small refund. Find more information on the Portland Bureau of Environmental Services website at **www.portlandonline.com/bes/index.cfm?c=43081**.

- **Texas tax incentives:** Texas residents can receive a rebate of any sales tax spent on rainwater harvesting and gray water materials. Such systems are also property tax exempt. Consult the Texas Manual on Rainwater Harvesting for full details, available in a PDF at **www.twdb.state.tx.us/publications/reports/RainwaterH arvestingManual_3rdedition.pdf**.

Find more information and useful links for a variety of state and local programs at:

- *Rainwater Harvesting for Drylands and Beyond* by Brad Lancaster, at **www.harvestingrainwater.com/rainwater-harvesting-inforesources/water-harvesting-tax-credits**, companion site to books on rainwater harvesting Brad Lancaster wrote.

- Harvest H2O (**www.harvesth2o.com/index.shtml**): Online water harvesting community.

- "The State of Rainwater Harvesting in the United States" (**www.rainharvest.com/more/StateRainwaterHarvesting US.pdf**): A magazine article about the evolution of rainwater harvesting practices in the United States and what the current trends are for the future from *On Tap*.

- Rainwater Harvesting and Gray Water Reuse Resources (**www.ci.tucson.az.us/water/docs/rain-gray-resources. pdf**): This is a comprehensive list of resources for more information on rainwater harvesting and gray water.

Incentive, rebate, and tax deduction programs are updated frequently and change from year to year. To find the most current information relating to your particular project and county, city, or state regulations about water harvesting practices, check with your local water district, your state's office of environmental quality, your city or county government offices, your county cooperative extension office, or even local builders groups to find out what types of incentives you might qualify for.

Ways To Store Water

The range of storage options and containers reaches from one end of the spectrum to the other. You can choose to store your water in stacks of 1-gallon jugs while your neighbor could construct a 10,000-gallon capacity tank. The size and design of your water storage depend on your budget, your needs, and your level of commitment to construction and maintenance. With many projects like this, you can start small and build on to your system as your skills grow.

Each household will have different water storage needs, depending on the household's size, daily water usage for cooking and cleaning, and possible emergency needs. For example, if you live in an area without fire hydrants, you might consider installing the largest water storage system possible so the fire department will have a water source if there is a fire.

The type of water you are collecting will also somewhat determine how the water should be stored. If you are stocking drinking water for emergencies, you should select easy-to-carry and sterile 5-gallon jugs. If you are redirecting rainwater for landscaping, you can select an open-air barrel. Many of the storage options listed can function for both types of storage, and there is overlap among the choices. After you have read through this book, you will have a better understanding of what works best for each system, and then you can make the decision as to what will fit your family's needs.

As a reminder, the following are some rules of thumb when planning the amount of water you need to store and the type of storage that will work best for your individual household:

- Determine the amount of water your household uses on a normal day.
- Decide your options for changing how you use water in your home.
- Determine the water required for your emergency food supply.

Determine whether you live in an area with special circumstances, such as areas prone to earthquakes and hurricanes. *Later chapters of this book will cover the systems possible for collecting water for household use, and sometimes these systems can also be employed for water supply sources before or in times of emergency.* However, many of these systems rely on electrical power — or at least a generator-driven pump — and these sources of power might not be available during an emergency. As discussed in Chapter 1, your home could be without power for days or weeks during a natural disaster. Additionally, storing water for emergencies should be done well in advance of when you need it, so your efforts can be focused on recovery and cleanup versus water collection. *For more information about planning how much water to store, refer back to Chapter 2.*

A way to begin is to start by storing enough water to get you through the recommended three days of an emergency. From this, you can build up your stores and add to your system as your budget, time, and expertise allow. As you become more aware and accomplished at water storage, you will be able to increase your emergency supplies in a manner that best fits your family and lifestyle. *The following sections will cover all sizes of storage options – pick and choose the methods that work best for you in creating your own emergency water supply.*

Small Storage Options

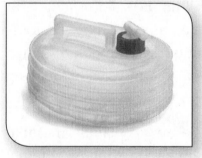

▶ Collapsible 5-gallon water jug

Your storage containers are one of the most important elements in your system. They will hold your water, sometimes for an extended time, so be sure they are secure and are the right size for the job. There are a multitude of options from which to choose, so you must decide which containers will fit your plan, climate, and budget.

Water tanks made for camping, traditional home emergency water storage containers, and rain barrels are all smaller storage options. For this book's purposes, the smallest containers discussed hold 5 gallons of water. Because you need a gallon of water a day to survive, as FEMA recommended, smaller containers would not be practical for long periods of time. However, if you have only small spaces for storage, you might want to opt for smaller 1- or

2-gallon jugs. This smaller jug also has the advantage of being more portable. A gallon of water weighs 8.3 pounds, which means a 5-gallon container will weigh about 41 pounds and might be hard to manage.

Finding small storage containers

You can find a variety of containers from camping and outdoor suppliers and restaurant supply companies. Your container needs will depend on how, where, and when you will store or use the water. It is not recommended to reuse plastic containers that have held other liquids, such as milk, juice, or oil. These cannot be properly sanitized and will not be safe for long-term storage. When considering which container to buy, consider its strength, weight, and storage ability. As you look for small water containers, consider the following features before making a large investment:

Food-grade plastic materials. These are containers meant to store food and keep it safe from contamination, from the external environment and from leeching of the container itself.

Stackability. A container that stacks will make storage much easier. A family of four needs a minimum of three 5-gallon containers for a three-day water supply. The containers will add up quickly, and you will want to be able to store them in the smallest space possible. It is important to choose containers strong enough to support the weight of the jugs stacked above them. Shape is a consideration, too because square containers are much easier to stack.

Opening size. Before using a water container, sanitize the inside with a solution of chlorine bleach and water. It is important to have a large enough container opening to accomplish this. To properly sanitize your containers, FEMA advises adding 1 tsp. of unscented regular bleach to 1 quart of water and swirling this throughout your container. Make sure to touch each surface. Pour out the bleach solution, and rinse the container with clean water. When this book mentions bleach, it refers to plain household bleach without scents or thickeners added. Bleach for colored laundry will not work either because it has special additives.

▸ Water storage barrels with spigots for easy pouring

Spigot or pouring spout. This might seem less important, but consider the strength of the person who will use the water containers. If someone who is not strong needs to pick up the full container and pour from it into a cup, this could be awkward and result in a loss of water. An easy-to-use spigot will solve this problem. Also, if the container is too heavy, the spigot will water easy to get without help.

Handle. A sturdy way of carrying is essential for an easy-to-use water container. Smaller containers are meant to be portable, so most will have a sturdy handle. This will make them easier to move if you need to evacuate your home due to an impending natural disaster or even to move about in your home when you need to use them. Storing water in these smaller, 5-gallon containers is a practical way to keep emergency water on hand.

What if I need a bit more storage?

If 5-gallon containers do not fit your storage needs, you can find similar containers in 7.5-gallon size, 15-gallon size, and 20-gallon size from many of the same manufacturers. Keep in mind the weight of water when you are choosing containers. One 20-gallon container might seem better than four 5-gallon containers, but it will weigh more than 160 pounds when full and will be difficult to move.

The next option is a 55-gallon barrel. You can store 55-gallon barrels of tap water, or consider using them as part of a rainwater harvesting system. A tank of 250 gallons or less will be considered small. Steel-reinforced, 250-gallon, food-grade tanks are not difficult to find. For a new tank, plan to spend about $1 per gallon of storage. Used tanks of this size are considerably cheaper but require careful cleaning, unless you want your water to taste like whatever was previously in the tank.

Cycle through your containers

No matter what small-size storage option you choose, you will need to cycle through the water on a regular basis. For 5-gallon containers, empty and refill them every six months to one year. FEMA recommends replacing your water every six months. Do not simply dump the contents down the drain and waste all that water — it is usable. Instead, incorporate the containers into your normal routine and use some of the water for cooking, cleaning, watering plants, or camping. After each container is emptied, cleaned, and sanitized, refill it and mark the date so you will know when to go through the process again.

For larger containers, such as the 55-gallon barrels or the 250-gallon tanks, you will also need to cycle through the water on a regular basis. Some manufacturers recommend emptying, cleaning, and refilling large holding tanks every couple years. Letting water sit too long might cause it to take on unpleasant flavors or go flat for lack of oxygen. This water is still drinkable but not tasty. If you decide to empty your tank and do not want to drink this water due to the taste, it can be used for watering the garden, filling a pool, or filling smaller containers for household use, such as filling toilet tanks or washing machines.

Large Storage Options

If you need larger amounts of water or have the space, you can use ponds, large tanks, soil-direct aquifers, open water tanks, and even swimming pools to store water. This book considers small containers to contain 250 gallons or less, so large containers hold more than 250 gallons. These emergency storage solutions are the same as those used for regular household water storage. Keep in mind, these are large systems that do require ongoing monitoring, consistent maintenance, and a hefty upfront expense. Large storage containers come in a variety of shapes, sizes, and materials. Corrugated metal, stainless steel, fiberglass, plastic, stone, and cement are some of the options available.

Common materials for large storage tanks

When installing a large storage tank, you have the option of buying the entire system pre-constructed, such as a large cistern, or you can build your own tank from the ground up, such as a cement-block above ground tank. Each of these systems has its distinct advantages — and disadvantages — so the choice depends on your needs and the region in which you live. You do not want a storage tank that will be damaged by freezing temperatures if you live in the North.

Once you make the decision to install a large tank, consult with an expert from your Soil and Water Conservation District (SWCD), a nearby farm supply store, or a large home improvement chain store. These sources can lead you to the best choice for your region and your budget. Choosing a local supplier will also save money on shipping, as these tanks are heavy, bulky, and expensive to ship. *As you start your research, the following sections will highlight a few of the most popular choices for tank construction and help you decide which is most appropriate for your needs. Following chapters will go into more detail regarding installation, usage, and maintenance of these various structures.*

Plastic: Plastic tanks are one of the simplest solutions — you can pick one up at most farm supply stores and take it home in a truck. It might need some modifications to work, such as properly spaced inlets and outlets (entrances and exits for the water) and some screening to prevent debris from coming in with the water. Prices vary depending on the size of the tank, but tanks that will hold more than 250 gallons cost 50 cents to $1 per gallon of storage. You also will have to buy a concrete pad or base for the tank on which to sit. Note that if you are storing water to use in a fire, do not use a plastic tank because it will melt at a high heat and the water will be gone before it can be used to fight the fire. *Storing water for fire suppression is covered in more detail in Chapter 8.*

Cement: Cement tanks are affordable, but if you build them yourself, you will complete a great deal of labor. The cost of a cement tank is more difficult to estimate because the size and structure can vary greatly. If you are doing the labor yourself, estimate about 25 cents per gallon or more. Cement tanks will last a long time and can be built in a variety of shapes and sizes to suit your needs. Cement will also withstand high temperatures and store water to extinguish fires.

Metal: Metal tanks come in either corrugated or flat metal; riveted, bolted, or welded; and galvanized or stainless surfaces. Corrugated, riveted water tanks can hold 20,000 gallons of water or more. Bolted steel tanks can hold from a few thousand gallons to 3 million gallons. Prices vary widely based on size, base requirement, and labor costs for bolted tanks. Most of these large steel tanks are used commercially, but you can use a smaller variety on your property. These are if you run an active farm.

Fiberglass: Fiberglass, which is rustproof, durable, and for underground storage, is another option for large water tanks. In addition to the price of the tank, you will also have to pay for shipping and placing the tanks, which will cost more than other types of tanks due to the weight. Using it underground will also take construction and might require outside help.

Swimming pools

If you have a swimming pool, you most likely did not construct it as a way to store water for emergencies. However, it will come in handy because even a small pool holds a significant amount of water, and it is already disinfected with chlorine. If you maintain a regular pool treatment schedule, fully change the water at least one to two times per year, and partially change the water after heavy use, and you can confirm the chlorine level is between three and five parts per million (PPM), the water might be safe for drinking. Chlorine is measured in ppm, and a typical swimming pool has between three and five ppm. For comparison, municipal water is also treated with chlorine and ranges from 0.2 to 3 ppm.

Although the pool water is treated with chemicals, it will not be safe to drink without additional purification. Chlorine is not the only chemical present in pool additives, and these extra chemicals will have to be removed before drinking. Additionally dirt, leaves, bugs, and other debris will be present in this water and should be filtered out before use. Filtered pool water is fine for non-drinking purposes, such as showering and cleaning, but if you plan to drink it, you must follow an additional purification method, such as boiling. *Chapter 6 covers these methods.*

Earthworks and natural structures: beyond barrels and tanks

Storage of water does not necessarily require filling a large tank. Using or collecting water efficiently as it moves through your property is also considered storage of water because you are helping the water make its way back into the aquifer where it is naturally stored. Even making sure rainfall is allowed to filter back into the soil is a way to store water. If you are using a well for your main water source, properly recharging this aquifer ensures you will have a water table beneath your well.

Using natural sources, such as channeled runoff or constructed earthen structures, such as ponds, to collect water is a solution for many areas, including those with excess rain or water scarcity. Earthworks, berms (mounds of earth with sloping sides located between areas of about the same elevation), and small dams to direct water flow on your property can ensure gardens, trees, and other plants receive all the water they need. These natural structures put the water to its best use and reduce erosion and waste during heavy rainfalls. Most of these systems, however, will require professional assistance due to issues, such as potential groundwater contamination. Inquire at your local SWCD for help in designing, building, and implementing an earthen structure for water retention. *A design for using earthworks in your rainwater collection system is in Appendix D.*

Berms

A berm is a manmade land feature that can refer to a few different land formations. Berms can be flat strips of land, such as a strip of grass next to a road; raised banks, such as a wall of earth built for defense; or a terrace next to a canal. The earthworks and berms can be as simple as a mound of mulch that slows and redirects the flow of water or as complicated as a series of low earthen walls to contain and redirect water on a property. Water will always flow to the lowest point possible, and this is key to using earthworks in water control and collection. Your goal is to slow the flow down enough so the plants and soil can absorb the water.

For example, if you have a large hill on your property, the water might run down quickly during a rainstorm. This creates ruts and gullies, washes the soil away, and makes it an unusable space. Furthermore, the water flowing through this space is not properly absorbed and put back into the groundwater. A line of large stones, a well-placed log or two, or a speed bump made of soil on the hill will slow the water down, reduce erosion and washout, and ensure more of the water soaks into the soil and surrounding area. Several of these obstructions, strategically placed, will slow water runoff, direct rainwater to trees or other plants that need it, and ensure the rainwater can eventually make its way back into the aquifer.

With a large enough space or a long enough spillway, you can also build a series of low dams to slow the water down. Just as in a fast-moving stream, low obstructions, such as a series of rocks, logs, or built-up earth, will stop the forward momentum of the water. By slowing this momentum, you are taking away some of the force of this water as it builds up speed flowing downhill.

A system of well-placed dams can easily slow down water in an area that has runoff problems, such as a drainage ditch or small creek. This situation ensures your property is absorbing the moisture it can.

Along with slowing water down, earthworks can direct water to the places where it can do the most good. If you plant a tree in a depression in your yard, it will naturally collect water any time it rains, and that tree will grow to be strong and healthy. Instead of just looking for those naturally occurring areas in your yard, you can create them yourself.

Ponds

If you need to store a great deal of water, and space is not an issue, a pond might be the best solution. When treated properly, pond water can be used like rainwater. However, planning a pond for water storage is not as simple as just digging a hole in the ground. A pond needs to be properly placed and planned to gather the largest amount of rainwater and avoid contamination. Check with local authorities about laws concerning ponds.

A grass-lined or -edged pond causes the least amount of sediment buildup in the water and is the cleanest possible option. Placing the pond in an area that will naturally catch water should keep it properly filled. If properly placed, it will harvest groundwater runoff the way gutters and rain barrels capture rainwater from a roof. For example, you might want to place your pond at the base of a hill. The pond also needs to be in a place where road runoff, livestock runoff, chemicals, and fertilizers will not contaminate the water.

▶ This pond was created at the base of the hill to collect rainwater used to water crops.

Water in a pond can be used for livestock or irrigation, or it can be filtered and purified for the household, with the addition of a pumping system and holding tank. Consider the eventual use of the pond when you plan its location. A full pond downhill from a house will require plenty of electricity to pump water up to the house. If you want to use gravity to feed water into the system, you will need to place the pond on higher ground. However, a pond placed on high ground will only fill from rainwater and not benefit from runoff on the ground. Ponds for water storage are most common in agricultural areas or places where there is an abundance of land available.

CAUTION! A large, open water storage area, such as a pond, is a drowning hazard for small children, animals, or even adults. If you introduce this type of structure to your property, take precautions to protect those who might wander too close to the water. Consider a fence around the pond, or string a line across the water someone could grab if they fell in.

Springs and wells

Using water from a natural spring or a well drilled into an aquifer is a great option for many households. The water must be filtered and occasionally tested for purity, but it is an accepted form of household water supply in rural or unincorporated areas. However, unless your well does not rely on electrical power to pump water, do not consider it an emergency water supply. If the well has a holding tank water is pumped into for household use, this tank can be used in an emergency. The size of the holding tank will determine the number of gallons available for emergency supply in a power outage.

A spring is an acceptable emergency water source if the water can easily be removed and either boiled or treated with chlorine bleach to kill bacteria. Water from the top of the spring will be cleaner and more aerated than water near the bottom and will taste better for drinking. Water near the bottom has a lower oxygen level and is more likely to contain large amounts of silt and sediment, decaying leaves, and other things that would cause the water to taste bad or contain unhealthy organisms.

CASE STUDY: DESIGNING A SYSTEM AROUND STRICT WATER COLLECTION CODES

Judith Walsh Project
18953 Viewcrest Drive
Sonoma, CA 95476

Architect, David Marlatt
161 Natoma St.
San Francisco, CA 94105
david@dnm-architect.com
www.dnm-architect.com
415-348-8910

The Walsh Project was a challenge for us on many levels. This family came to us with these goals in mind:

1. To have more water available, which seems increasingly likely in California, given climate change impacts.

2. To help minimize drawdown of the local water table, because they are on a private community well system likely to experience increasing demands as the population grows.

3. To not waste the water that falls on the 3,000-square-foot roof in winter rains.

4. As a fallback.

They did not approach this as a way to save money because they knew it would take a long time to recoup the initial investment. Unless water rates rise drastically, this sort of system will never pay for itself. No one should install a rainwater harvesting system for purely financial reasons. The hope is, though, in the future as water becomes more scarce, the economics might better support the storing and harvesting of water.

This was our first project with water storage as its main goal, and through our research, we learned about best practices and regulations. We found that California has placed so many regulations on gray water that it did not make sense to use that as a source for any of our water storage. We ultimately concluded that gray water would also be too costly because we would have to install four separate systems (regular sewage, site runoff, roof rainwater, and gray water). Because we can capture and store more rainwater than gray water in the household, we opted for the rainwater system. The disadvantage of this is that the rain is captured seasonally when it is needed the least, but this challenge is addressed by the cisterns.

▶ Walsh's water tanks. Photo courtesy of David Marlatt.

We wanted to use four 5,000-gallon above-ground water storage tanks with the goal of storing up to 20,000 gallons of water. These tanks would be located on a hillside below the house, so we used electricity from solar roof panels to pump the water up the hill. Our civil engineer had to revise the drainage plan to divert rainwater coming from the roof into the cisterns before it mixed with the site's runoff. This is to avoid silt buildup in the irrigation system. Our current plan is to use harvested rainwater to significantly reduce irrigation water needs because the current plumbing code does not allow stored rainwater to be used for household purposes.

The biggest drawback to this system is the cost of installation. It took many experts to design this system, in addition to the material costs and my architect services, we consulted a water catchment specialist, a civil engineer, and a grading contractor. Second, the visual impact of the cisterns is not the loveliest feature in the landscape. Above-ground cisterns, however, were about four times cheaper than buried cisterns when you include the cost of excavation and added installation charges. We are also looking at installing a first-flush device to reduce the impurities deposited from the roof over the summer dry system.

We are still hoping that times will change and states, such as California, will see the wisdom is harvesting rainwater more aggressively. Right now, codes do not permit using stored rainwater for interior household use. I would argue that well water does not constitute stored water because pulling water from a well can deplete the water table more quickly than the winter rains build it up. It is going to take continued pressure from concerned citizens across the country to change these codes.

Do You Need Footings and Foundation?

For larger projects that hold water, such as cisterns and holding tanks, your structure will need to be supported or built on a foundation. Your home, garage, and outbuildings are all set on top of a foundation because a sturdy, long-lasting building begins at the bottom. This is especially true when constructing a tank that will hold a large amount of water. Water weighs 8.3 pounds per gallon, so a 400-gallon tank will contain more than 3,300 pounds in water weight. Add to this the weight of the walls, and you can see that even a small holding tank exerts downward pressure on the soil. The foundation under your tank will help distribute this weight evenly across the ground surface so the structure will not slowly sink into the ground over time or shift from side to side if the ground settles or heaves from freezing and thawing. This sinking or shifting could cause the walls of your tank to buckle, crack, and eventually fail and cause the tank to break apart. Movement of your tank could also pull on the pipes coming in and out and cause them to break apart and fail.

▸ Here, a concrete foundation is being poured for a large water storage tank.

Cement or concrete is most commonly used to build footings and foundations. The type and dimensions of footings you need vary based on your soil type, your region, and the size of tank or cistern you are constructing. Footings must always be built below the frostline, so groundfreeze does not affect them, and they do not thaw each season. Building the footing forms and pouring concrete are a science in themselves, and entire books have been written covering this field. Laying in a foundation requires some skill in cement work but can be accomplished by most do-it-yourselfers. If this is your first concrete or footings project, it is highly recommended that you consult with an experienced friend or check out one of these complete masonry books before starting:

- *Building with Masonry: Brick, Block and Concrete.* Written by Richard Kreh, published by Taunton Press in 1998.

- *Residential Construction Academy: Masonry, Brick, and Block Construction.* Written by Robert B. Ham, published by Thomson Delmar Learning in 2008.

Basic procedure for installing a foundation

The first step in putting in the footings is to mark off the dimensions of your tank. When you are done with these measurements, you will have the perimeter of your structure delineated with stakes and roping. From this perimeter, you will dig, form, and pour your footings and foundation. Keep in mind that the foundation will cover up these measurements, so you will have to re-measure if you are planning to build your tank from concrete. If your tank is going to be placed below ground level, you will have to excavate the site before you can mark out the footings.

To measure for a circular structure: Place a stake where the outside edge will be — the best place to start is where the pipe will enter or exit the tank because this is an important point to keep constant. Place a stake at this point, tie a cord or rope to this stake, and extend it out to where the opposite side of your tank is. Place another stake there and tie the cord to it; this line represents the diameter of your tank. Find the halfway point of this line, and drive a stake in the ground at this mark. Cut another piece of cord equal to half the diameter of your tank, and tie this cord to the stake placed in the middle. Use this line as a homemade compass to mark the outside edge of your tank — wherever the end of the line reaches will be the perimeter of the tank. Place stakes every 2 feet. As you place each stake, be sure to mark it with information regarding its location, such as "entrance pipe, N corner." When all the stakes are in place, string a cord around them to form the perimeter edge of your tank. Again, this is only to give guideline on where to put your footings, and it does not have to be measured to the inch.

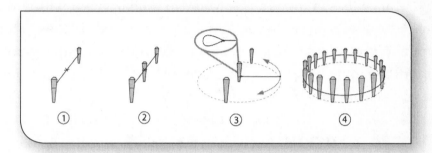

To measure for a square or rectangular structure: Use the same cord and stake method as when you measure for a circular structure, and make sure to place the first stake at the point in your structure where the pipes enter and exit. Be sure to mark each stake with specific information, such as "SE corner, exit pipe." From these stakes, measure out to the corners and place stakes at each corner of the structure. Run a cord from stake to stake, forming the perimeter of your tank. At this point, you can check to make sure the corners of your footing line up at 90-degree angles or are "square." To measure for a square, measure the distance diagonally from each corner stake — the distance should be exactly the same from corner to corner. If it is not equal, adjust the stakes and measure again. This step is not essential for placing footings and foundation but helps you later on because you will have an evenly laid-out base. For the purposes of marking your footing area, you only need an outline of where the structure will sit.

From this cord, you will be able to determine where your footing trench will be. The walls of your tank or structure will sit in the center of these footings, so the footing trench must extend around the entire perimeter. The trench will be 16 inches wide and at least 9 inches deep. The dimensions of your footings are determined by local code, so now is a great time to double check with your inspector and make sure you are pouring the correct footings for your type of project. You will also need an inspector to sign off after the footings and foundation are poured — and before you install your rain barrel.

CAUTION! Always call the utility companies before you dig. Who knows what lies beneath the soil in your yard? Accidentally hitting a gas, electric, phone, or cable line can be costly and dangerous. Avoid any problems and call your local utilities before you start any excavation. They will come out for free and mark all the lines of which you need to be aware. Most communities have a one-call hotline that will alert all utilities; just ask at your city offices for more information. Most utilities need at least 48 hours advance notice to get their crews to your site.

As you dig the footing trench, keep the bottom of the trench level, but leave the soil as untouched as possible — only dig where needed so the remaining soil stays intact. Undisturbed soil will be much more compacted and stable than soil that has already been dug up. Build the cement forms according to the your local code and by standard construction practices. As you build your forms, check to make sure they are level, and adjust as necessary. After all your forms are placed, you are ready to pour the cement. For larger projects, consider having premixed concrete delivered to your site. This is more expensive and time-sensitive, and it requires more helpers during the pouring process. Mixing your own concrete is perfectly suitable for this stage but will take more time to finish as you mix each bag of concrete.

After the concrete is poured, be sure to give it the full curing period or hardening time the manufacturer recommended. Curing refers to the time immediately after the concrete is poured and finished; during this time, the concrete must be kept slightly moist and protected from sunlight and top pressure, such as foot traffic. It can take up to 14 days, depending on the temperatures and rainfall. It is best to refer to the specific instructions included with your brand of concrete. Do not rush the curing process; this is an important step and will create concrete with maximum strength and durability. Also, if heavy rain or unusually high temperatures threaten during the curing period, keep the concrete covered with a light-colored tarp. *Step-by-step instructions for building a simple foundation are included in Appendix D, Plan No. 1.*

Adding a drainage tile

It seems a bit counterintuitive to channel water away from your foundation, but if you are building or installing an underground holding tank, you will need to consider putting in a drainage line or drain tile around your structure. Once you introduce a new hole to your topography, it will attract water, and you might encounter unexpected drainage problems. Water draining in or near the tank's walls is a problem because it might pool and seep into the structure and introduce contaminants to your stored water. Additionally, water sitting around the buried exterior of your tank can exert pressure on the walls when it freezes and potentially cause your tank to crack, fail, and collapse.

TIP! Drainage "tiles" are a bit of a misnomer because they are not tiles. They are continuous, flexible plastic pipes used to channel unwanted water in a certain direction — away from structures or growing areas.

The only way to avoid this possibility is to install drainage tile around the footings — and halfway up the hole if you have extra drainage concerns. This is a simple and relatively inexpensive step that will give you added insurance, and putting drain tiles in from the beginning is much easier than digging out your tank when you find drainage issues.

Lower drainage tiles are best placed after the footings are cured, and the best system is to use 4-inch-diameter perforated plastic drainage tiles wrapped with a sock to prevent soil infiltration into the tile line. You can find pre-wrapped drainage tiles at farm supply stores, landscaping shops, or home improvement centers. Before placing the tile, pour in a 2-inch layer of gravel around the perimeter of your footings. Slope the gravel slightly toward the lowest point of your structure. Lay the tile on the gravel so the top of the drain tile is just below the top of the footing. If using more than one piece of drain tile, connect and seal each section according to the manufacturer's instructions, and finish with the open ends about 1 foot past the front corners of your foundation. As you place the drain tiles, make sure no dirt gets inside the line or it will become clogged and no longer work.

▶ Example of a French drain outlet

Where the tiles end, dig a hole about 3 feet wide by 3 feet deep, and fill it with gravel. Place the open end of the line in the center of each filled hole. Some people additionally run a non-perforated pipe, such as PVC pipe, out of this small pit, so the water is carried even farther away from the structure and collected in a gravel lined pit or in a holding tank. This is often referred to as a "French drain" and will allow you to collect this water for later usage. After installing the line, cover it completely with about 12 inches of crushed stone or gravel to help the water seep through to the drainage line and support the weight of the backfill. While working with drainage tile, be careful not to step on it or crush the line.

Installing a Gray Water System

Installing a well-functioning gray water system begins with assessing your property's features, outlining your goals, and determining the potential gray water volume you can expect. You will also need to do further investigations of the legality and regulations specific to your area. You might be the groundbreaker here, and your system could be the first one of its kind in your neighborhood. Expect a few questioning glances, and be prepared to thoroughly explain every aspect of your project. This chapter discusses the typical process of planning, installing, and using a gray water system. The following sections will cover the nuts and bolts of putting in your own system. *Appendix D shows a few plans for building a simple gray water system.* A do-it-yourselfer can perform — and use and maintain — many of the projects. It is highly recommended, however, to consult with an expert when it comes to working with your home's plumbing. This step might be required for the inspector to sign off on your permit.

Gray Water Health Concerns

The biggest concern with using gray water is the potential for contamination of existing clean drinking water stores or supplies. Gray water is never safe for drinking (people or pets), dishwashing, food preparation and cooking, or for irrigating food crops that are eaten raw, such as lettuce. Even the "cleanest" gray water will contain soil and soap from laundry, dish soap and chemicals from your appliances, and even pathogens removed from your body or bed linens. This water will not hurt your plants or soil when applied directly, but running it to an area where it might flow into drinking water stores is dangerous. This runoff can quickly pollute your drinking water. There is no affordable, reliable way to thoroughly purify gray water, especially as a homeowner. Even in the most dire emergency situations, drinking gray water is not worth the health risk. Some large cities experiencing extreme water shortages are working on methods for purifying gray water, but this technology is still in development.

The ultimate purpose with reusing gray water is to use this water to its maximum potential and to eventually return this "slightly used" water to the soil and allow it to be naturally filtered back into the aquifer. Gray water does not need to be purified when used for irrigation or within plumbing systems, such as a toilet tank. You might have to slightly filter your gray water to remove large pieces of debris so your plumbing or hoses do not become clogged as the gray water passes through. Large-scale gray water systems, such as those using drainfields or grass strips, also filter the water, but a professional best designs it to ensure the gray water flows correctly and reaches the proper destination without adverse runoff issues. *Simple filtering methods are also covered in*

Chapter 6. When gray water is directed and use correctly, there is no real health risk to drinking water. In addition to the runoff issues covered previously, try to limit human contact with the water as much as possible. Here are some ways you can prevent possible contamination issues:

⊕ Label your outdoor faucets and plumbing to let people know which is fresh water and which is gray water. Mark hoses as well, and only use the gray water hose for gray water usage to prevent cross-contamination. When you use the gray water hose, use gloves and wash your hands with fresh water when you are done.

⊕ Do not use gray water on plants you plan to eat raw from a garden.

⊕ Gray water systems work best when you are using water relatively clean to begin with. Avoid using gray water from soiled linens or cloth diapers.

⊕ Do not overload the system. Have a diverter switch that allows you to direct water entirely to the sewer line if needed. If there is going to be heavy usage of showers, laundry, or any other gray water source, you might want to switch to the sewer line so it does not overflow your gray water system's capacity.

⊕ If you are using heavy-duty cleaners or other harsh and potentially toxic chemicals that will be passing through your gray water system, you might want to divert back to the sewer line. The best gray water is relatively clean and free of potentially toxic substances.

Researching laws and regulations

The first step in installing a gray water system is to research laws concerning gray water systems in your area. A place to start is to use a search engine, such as Google, **www.google.com**, and type "laws gray water systems (your state name)." An example of this type of search using "North Carolina" yielded results containing a document titled "Untreated Gray Water in North Carolina is Waste water/Sewage." This document contains a number of laws and regulations about gray water and its legality in North Carolina. The document says that gray water can only be disposed of in a municipal system or a septic tank. Be sure to read through your results thoroughly because there might be exceptions, such as this one found for North Carolina:

"According to the 2006 N.C. Plumbing Code, treated household gray water might be permitted for use for specific purposes if treated according to Code Standards. In Appendix C, Section C101.1 allows for recycled gray water to be used for flushing of toilets that are located in the same building as the gray water recycling systems. These recycling systems can also be used for irrigation purposes when approved by the authority having jurisdiction. Appendix C includes information regarding the installation, filtration, disinfection, drainage, and identification of gray water recycling systems. Gray water used in a gray water recycling system must be filtered and disinfected before it can be recycled for flushing of toilets or irrigation as stated in Appendix C."

According to this document, with permission, the water can be reused to flush toilets and used for irrigation. When searching for information online, be sure to look at the date the information was published, as it could be outdated. Look for the most current information, especially if the site's postings are more than 1 year old — rules might have changed since this information was posted. Be sure to select information from government sources (sites that end with .gov) or reputable university sources (sites that end in .edu) because these sites are more reliable and accurate than other sites.

After your Internet research, it is also an idea to make a few phone calls to governing boards in your state, city, or county. A few examples of places to start would be the State Building Code Council, the state's Department of Insurance, the Soil and Water Conservation District for your area, or the local Cooperative Extension office. Expect to explain your plan — many people are going to be unfamiliar with what you would like to do. Once you have found the person who can help you, consider asking the following questions:

- What are the laws and regulations regarding gray water systems?

- What permits are needed, how much do they cost, and where can you apply for them?

- Will the area need to be inspected? If so, by whom and how can you contact them?

- What other agencies will you need to contact?

- Do they know of other people who have set up gray water systems in the area? How did they do it? What is their contact information, or can they pass on a message to make contact with them?

- What are the challenges others have faced with installing a gray water system?

- Are there any tax credits or incentives for installing a gray water system?

If, in the end you find that it is not legal to employ a gray water system in your area, you will have to put your project on hold. In the meantime, you can work to change the laws by lobbying lawmakers in your area. Until the laws change, you will have to make do with basic water-saving systems and conservation efforts.

Goals and Common Considerations

Once you can legally build a gray water system and have an idea of the laws, regulations, and permits needed, it is time to consider your goals for creating a gray water system. *Appendix C contains an assessment worksheet you can use while you answer the following questions.* The worksheet is a way to organize your ideas on the size and scope of your gray water project, its feasibility, and how much your project might cost. You must weigh the pros and cons of a gray water system long before you decide to build one. There are numerous possibilities for developing your personal gray water — from a simple pipe extending from your dishwasher to a full-fledged holding pond to collect all your household output. The following is a list of questions as a starting point to help you determine what will work for you. As you read through the following chapters, refer back to these questions and thoughtfully consider each aspect before you commit to this type of project.

⊕ **Are you willing to change your lifestyle to include a gray water system?** A gray water system is not a small change; it is a serious lifestyle change. Running a household with gray water is more time-consuming than turning on the tap, and deciding your level of commitment is also necessary when you take on this sort of project. It will take work and maintenance to ensure it is functioning properly. *More information concerning the upkeep and usage will be discussed in Chapters 6, 7, and 8.*

⊕ **Is your gray water system being used to improve sanitation or reduce toxins in the environment through natural filtering?** What are your other goals in creating a gray water system? For example, other goals could be to reuse gray water for irrigation or outdoor features, such as a shower. A simple intention would be to just dispose of the gray water safely. These practical goals are your decision, and the reasons will be better than dumping gray water into a septic tank or the municipal waste water system. However, disposing of gray water safely will not decrease your need to use fresh water as much as reusing the gray water will.

⊕ **Do you have enough space to set up a proper gray water system?** Depending on the design you choose, you might need a cistern, a large drainfield, or multiple external overflow tanks to collect your gray water. You might also affect have nearby neighbors with your gray water outflow. Consider these space needs as you work through the design. If you find space concerns, look into other ways to store water, such as rain barrels or underground systems. Trying to install a gray water system where there is insufficient room is a major flaw in your water storage plan.

- ⊕ **Does the gray water system fit into the existing landscape at your home, or will you need to make some changes?** Is the goal of your landscaping to look beautiful or produce food? Some other possibilities include controlling erosion, working as a firebreak, creating privacy, stabilizing the slope in the property, or using it as a place to meet and relax outside.

- ⊕ **What sources of gray water do you intend to divert, and what will the purpose of the water be?** Will you only collect water from the shower, or will you use water from the kitchen sink as well? Will the gray water be used to water your yard, or will it be used in a flower garden? How will it fit into the look of your community and your neighborhood? For example, if you live in a historic district comprising 100-year-old homes that have been painted and landscaped a certain way, how would a gray water cistern in your front yard look in this type of neighborhood? The answers to these questions will determine how large a system you will build and where the system will be located on your property.

- ⊕ **Is your plumbing encased in the concrete of your foundation?** If so, a gray water system might not be the most cost-effective system to install because you would have to drill into the foundation to access the necessary pipes to divert water. You might still be able to access the runoff from your washing machine, but a fully functional system might not be possible.

⊕ **What are the health risks involved with a gray water system?** The amount of pathogens that can grow in gray water systems is about the same as water commonly used for irrigation, and it does not pose a significant risk to a person's health if used as irrigation water. It is not recommended that gray water be used as drinking water, however because it would require a significant amount of filtration to be considered safe.

⊕ **What is the permeability of your soil?** Your soil might not be suitable for a gray water system. If it is impermeable or even too permeable, it might not be possible to install a gray water system that would work properly. If your soil is too permeable, the water will move too fast through the soil and will not be properly filtered. If your soil is impermeable, the water will stand in the form of puddles and will not properly soak into the ground. However, there are ways around these problems, so do not let soil type be an absolute deterrent to your gray water system dreams. *The following section will discuss solutions.*

⊕ **Do you have the right climate for a gray water system?** Sometimes an area's climate can be inhospitable to a gray water system. In areas where the climate is wet, a gray water system might not be practical because it can create swampy conditions that attract mosquitoes, encourage the growth of unsafe microorganisms and develop an unpleasant smell due to stagnation and decay. In areas that are extremely cold, freezing temperatures might make it impossible to use a gray water system, especially if there are long winter seasons. The water would freeze on the surface of the soil and never properly filter through the soil.

◉ **Are you prepared to deal with any legalities you discovered in your research?** For example, permits and inspections can take up time and effort on your part. Make sure you are ready for inspectors on your property and long lines at permit offices if these are considerations in your area.

◉ **What is your budget for a gray water system?** Sometimes the creation of a gray water system might be cost prohibitive. Before you start digging and cutting into your plumbing, determine the costs. Also, account for the costs of permits, if they are required in your area, before you begin creating your gray water system. On average, gray water systems cost between $2,000 and $5,000 to install. This depends upon the type of system you choose, how much plumbing needs to be diverted, and how far you want the water to travel from its source to the point where you plan to use the gray water around your property. This is just an average cost, and the actual cost depends on your particular situation.

◉ **Do you intend to get a financial return for installing your gray water system?** Is the return attached to reducing the load on your septic system, thus reducing the need for repairs, septic pumping, or replacement?

The first step is to determine the sources of gray water on your property. A gray water system is set up by diverting water that would normally flow into a septic system or a municipal waste water system. This diverted water flow can then be directed to numerous areas — such as directly piping it into a holding tank for later use, leading a pipe out the side of the house directly to a waiting landscape, or guiding the water to flow into a wide drainage area so it can slowly soak into the soil and purify as it makes its way back into the water table. Your gray water collection can also be as simple as catching the water in a bucket as you shower and using this water to fill your toilet tank. There are hundreds of uses and combinations possible once you divert or collect gray water, and the more accustomed to living with gray water you become, the more places you will find to use this water.

? The Simplest Gray Water System: A Bucket

If you are not interested in diverting the gray water from your plumbing, you can simply capture it by using a bucket. This is the most inexpensive method but does require more labor and time to collect and move the water to where you will use it. As you wash dishes, shower, or use the bathroom sink, keep a bucket or rubberized wash basin underneath the tap to collect the used water. The one advantage of this method is you can avoid collecting dirty water by moving the bucket. Once your bucket is full, dump it directly on your plants. (Again, do not use this water on edible plants.) This system is just as efficient and effective as any complicated system, but it requires hauling buckets of water.

Right now, the plan for your gray water system begins with determining how much water you expect to collect, where you will capture the water, and where you will use the water. Again, your budget, desire, and time constraints for future use only limit the choices. You can also start small and build onto your system as you get used to using gray water. There are a few important plumbing considerations you must tackle before you can start designing your pipe work. These are not complex but do require a bit of explanation.

From where does your gray water come?

No matter how your gray water system is designed, it will be fed by water that originates from either a pressurized source, such as a washing machine, or a gravity-fed source, such as the pipe coming out of your kitchen sink. The type of system you are tapping into will affect how you run your system. These each have their advantages and disadvantages, and as you design your system, choose the type or combination of types that will benefit your situation the most and give you the best results. *The following section will explain the differences between these two sources. Appendix D also includes a few simple plans working off these two types of gray water feeds.*

Pressurized gray water

Pressurized sources of gray water come from appliances that pump and pressurize water to work the device and to empty water out after use. Examples of these are washing machines and dishwashers. Pressurized sources require a much simpler gray water system because they do not require gravity to move the water. If the distance is short, the water can be pushed up a slope or even vertically.

Diverting water from some appliances can be as simple as attaching a hose to the waste water outlet on your washing machine or dishwasher and running it out through a window or drilling a hole in the wall. This is a simple system, with minimum fuss, but it will only work when these devices are running.

An example of this system would be a "Laundry Drum Surge Tank." This system is simple to install and requires minimum plumbing installation. Here is how it works: As the washing machine water is drained, it surges into a large 30- to 55-gallon food-grade outside tank raised off the ground. This height gives the water some pressure, so you can attach a hose to the drum for watering plants, grass, or trees. You can add a water pump on the hose to increase the pressure, but this is often not needed. This system allows you control over where the water is going, but the water should be drained from the barrel within 24 hours to prevent the growth of undesirable microbes. *Please refer to Appendix D, Plan No. 2 and No. 3 for plans of building a laundry gray water harvesting system.*

CAUTION! If you decide to use a "hose out the window," you must always be sure this hose is placed into the holding tank before you run a load of laundry. If you forget, you could end up with an entire tub full of wash water pouring out onto your laundry room floor. Create a reminder for yourself — and anyone else who might do laundry — such as wrapping the hose around the laundry detergent or some other simple device, so you never forget to place the hose outside. Eventually this will become second nature to you as you do laundry, but at first it is better to be safe than sorry.

These laundry gray water systems can also be adapted for use with a dishwasher because this drains similar to a washing machine because the water is pressurized and is pumped out of the appliance. Be aware that a dishwasher water pump is even smaller than a washing machine, so it cannot push the water as far and will burn out faster if there is too much back pressure. This means that your gray water destination needs to be closer to the house, and you cannot use as many branches (no more than three) when setting up a dishwasher.

Gravity-fed gray water

Gravity-fed sources of gray water in your home include sinks, tubs, showers, and any other source that does not rely on a pump or pressurization of water. These fixtures use gravity to drain gray water and move it through the pipework and eventually out to the main city line or into your septic tank. For a gravity-fed system to work, there must be a fall in the pipe because your septic tank or waste water line is underground or just below your home's lowest level. The fall in your plumbing system uses gravity to build up enough water velocity to continue moving it as the pipework becomes more vertical. Without built-up force or a mechanical aid, such as pump, water cannot be efficiently diverted vertically for long distances. This fall and gravity pressure must also be at a downward slope for plumbing to work with gray water collection. Because of this "slope" necessity, gravity-fed systems require more effort and design to work efficiently. You will most likely need to install a pumping mechanism to get the gray water to where you are planning to use it. *The following sections contain more detailed information on fall, slope, and multi-flow or combined-flow lines.*

Most gravity-fed systems are designed as multi-flow systems because the water is diverted directly from the source. More complicated combined flow systems use these same principles but require the assistance of a licensed plumber because of the necessity of cutting pipes and diverting household pipework. This is not recommended for someone who is not a plumber; in some areas of the country, it might not even be legal for an amateur to rearrange plumbing on that scale. More important, you can create a mess of your household's plumbing if you do not know what you are doing, which can lead to water damage or expensive fixes. Furthermore, most of your water-dependent appliances could be damaged if the water supply or drainage is not adequate. You also need a professional to help you separate lines for gray water and those containing black water from toilets.

One example of an easy-to-build, gravity-fed system is a direct-to-landscape project, where you create the gray water outside your home's plumbing system and position this source in an area where the water will naturally flow into the soil and nearby plants. The most common type of this system is an outdoor shower. You can build your outdoor shower as simply or elaborately as you like.

TIP! Gray water systems work even better when the input is earth-friendly. Switch your family's soap, shampoo, detergents, and cleaning supplies to those made with natural, biodegradable ingredients. These can be found at outdoor stores, organic food co-ops, or made at home with vinegar and baking soda.

Designs range from a simple hose with a shower head attached to a fully built outdoor "shower room" with hot and cold running water. The goal is to have your shower water run through the base of your shower and gently flow to the plants needing the water. For this reason, build your outdoor shower near the plants or trees you want to water — or landscape around the shower with new plants. The floor of your shower should be on a patch of ground slightly sloped and will drain quickly. Clay is best, but you could also put in a layer of gravel or coarser rock underneath your shower. For comfort, cover the rock with a slotted pallet that will not be slippery when you are showering. Create enough wall and shelf space so you can easily store your towels and showering supplies. With a little creativity, you can create an outdoor grotto in which to shower, complete with privacy and beauty surrounding you.

Another energy-friendly solution is a solar shower that uses large bags of water hung in the sun to warm up the water. There are a number of online sources where you can buy solar showers, but you might also find them in a store that sells camping supplies. If you live in an area where the temperatures dip in the winter, your outdoor shower will only be usable for a few months out of the year. This is still acceptable, though because that is when your plants need water. *Appendix B lists resources for solar shower suppliers.*

✓ **Emergency Plan Tip!** An outdoor shower fits well into your emergency plan, too. During an emergency, you will most likely be doing dirty work but not have access to hot water. Keep a solar shower kit on hand, and you will always be able to clean up at the end of a long day.

Another slightly more complicated gravity-fed design is to simply divert the water from your sink's drainpipe right out the side of your house. *Appendix D, Plan No. 6 shows a simple plan for this design.* This system allows you to collect and redirect the water you drain from the sink and use it to water outside plants or the lawn. You only want gray water here, so a kitchen sink is your best choice. If you live in a temperate zone where it does not freeze, you will be able to leave this direct system hooked up year-round. However, if you live in an area where it freezes, you will have to switch this diversion on and off with the change of the seasons or reroute your diversion to an indoor holding tank. If you choose to turn it off during the winter, also drain the exterior pipe so it does not freeze and break apart.

This idea of rerouting pipes — with assistance from an expert plumber — can be adapted to work with nearly all of your drains and used for the majority of your landscape watering needs. Additionally, even if you do not need to water your plants, these gray water diversions can be advantageous to replenishing the groundwater by creating a simple drain field.

Determining Output Lines and Assessing Your Plumbing

Once you have decided what type or types of gray water you will be diverting, you will then have to determine how and where you would like to reroute this water. This determination begins at the end because you will need to decide whether to have a one main gray water line (called combined flow) coming out of your home or have multiple lines (called multi-flow) exit at various points. Again, each has its own benefits, and they can be combined or used as complements to each other. The easiest way to begin is with a multi-flow because this will take the least amount of work and is the easiest to add on to or maintain. *The following sections will delve into these issues along with more details on assessing your plumbing.*

Combined flow versus multi-flow

A combined flow system starts at every gray water source in your home, and lines are run from these sources to one main pipe. This pipe is then directed out of your house through one outlet to either the holding tank or a series of pipes from which you will use the water. A multi-flow system works off one or two sources of closely connected water, such as your sink and dishwasher, and the gray water is diverted directly from this source out through one pipe. In a multi-flow system, the gray water is not mixed with gray water from other sources. Most people use multi-flow systems; however, there are those who want a more water-efficient system that collects all the gray water from their home to be used in one location for a specific purpose. Knowing which system you will use will help you determine what type of plumbing to install.

In relation to plumbing, the multiple flow system is simplest because each gray water source can flow separately to its final destination, so minimal plumbing is required to access these waters. With a combined flow, the gray water is directed throughout the home to a centralized pipe and then sent out of the home through another single pipe. This requires more advanced plumbing and reconfigurations within your existing pipe system.

Multi-flow systems also take better advantage of gravity. Water flows down due to gravity, and even the oldest household systems put this to work to move the water through the pipes and out to the sewer. If you ever have to clean out a clogged drain, a "drop" in the line is essential for building up enough pressure to keep things moving. This same principle applies to movement of gray water. Visualize the path that the gray water from your upstairs shower needs to take to get to your flowerbed — the steeper its path, the faster it will get outside. This is why multi-flow systems work better with gravity: By only having one or two sources come together in a single pipe, the drop can be steeper. With a combined system, many vertical lines come together, making it more difficult to create enough drop to keep the water moving. Furthermore, once this combined flow reaches the central pipe, it still needs to be high enough for the water to flow even further downward to its final destination. So in a combined flow system, all of the pipes must be higher than the central pipe.

To decide which flow system works best for you, review the following sections highlighting the components of each. You can make your own hybrid of the two and create your own unique system based on your needs.

Advantages of a multi-flow system:

1. You have a greater range of places that the system can water. For instance, a bathroom can flow to the lawn, while water from the kitchen can flow to bushes in front of the house, and the water from the washing machine can be used in a flowerbed.

2. If you have a large house or one spread over a large area, then a multi-flow system might be a better choice. In some instances, moving gray water from one side of a large house to another might be difficult or impossible without a multi-flow system.

3. If you have a flat property, there might not be enough height for a combined flow system.

4. If your house has more than one story and some of the areas you wish to irrigate are above the first floor, a multi-flow system might be your only solid option because you would have to use the plumbing upstairs to irrigate some areas, and you would have to plumb the downstairs gray water sources to areas at ground level or below. Gray water sources on the first floor could not irrigate places above that floor due to gravity issues.

5. If there is little space between the height of your gray water sources and the areas you want the gray water to reach, you will not have a choice because a single flow

system requires a significant amount of what is referred to as fall, which is the distance between the source of water and where it reaches the ground.

Advantages of a combined, single-flow system:

1. If there are plants or a lawn in one section of your yard and you wish to water it specifically with gray water, a combined system will work better because all of your water will come out from one outlet. This could be because you live in an area prone to drought, or the location of your plants makes it difficult for rainwater to reach.

2. A multi-flow system because it has multiple flows, does not make for the most efficient method of watering because you will have to manage multiple output areas. For instance, if you wish to water your lawn using a gray water system, a multi-flow system would require multiple hoses extending out to your yard. It is still possible to use a multi-flow for lawn-watering, just more time-intensive.

3. If you need cleaner water and have to add a pump or filter, a combined system is much more cost-effective because you only have to buy one pump or one filter, as opposed to using a different pump or filter for every flow of gray water. All the gray water is collected at one central location and is then pumped and filtered before use.

4. You can choose where your gray water goes into a combined system rather than having to deal with multiple flows. You can move your hose to any area in your yard because some areas might require different amounts of

water at different times, depending on rain fall and ground permeability. With a combined flow, you have complete control over where the water will flow.

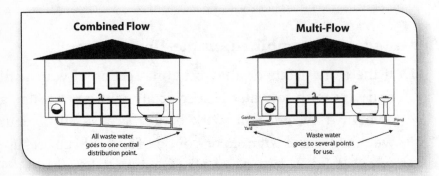

Once you decide what type of system, combined or multi-flow, you will begin to figure out the configuration of plumbing and where you need to access pipes. The more complex your gray water system is, the more plumbing equipment and labor you will need to put it together. If you only intend to access the waste water from your washing machine, your work will be simple compared to accessing the pipes to all the sinks and showers in your home. The more complicated the system, the more imperative it will be to hire a plumber. They will not only be able to help you access pipes and create your system, but they will also know more about the actual materials you will need to use.

CAUTION! This book is not intended to replace a professional plumber's assistance. In some areas, a licensed plumber's professional installation might be required for the system to meet certain standards and pass inspection.

Conserving fall or slope of the pipe

While reviewing your plumbing situation, be sure to take into account the fall, which is the vertical distance between the gray water source and the destination of the water. The flow of water in one direction creates a flow. In pipes, a slope of no more than 2 percent is essential for a gray water system to properly operate. A 2-percent slope equals a fall or drop of about ¼ inch per foot, or 2 feet across a 100-foot distance.

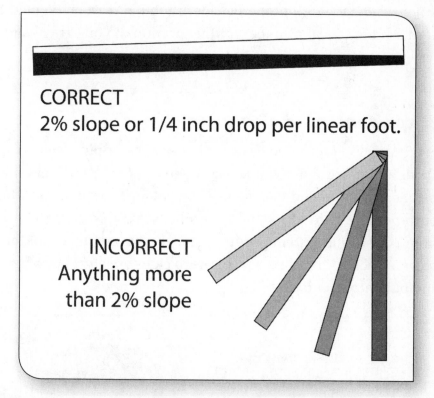

A steeper slope will allow water to flow too quickly and leave solids behind. A lesser slope will not carry the solids away, which might create a blockage. If you cannot create the proper slope to your pipe, make sure you or the plumber installs an adequate cleanout access wherever buildup might occur.

Remind your plumber that he or she needs to conserve as much fall as possible. Most plumbers are used to creating water systems that empty into a septic system or waste water line, which will be much deeper underground than a gray water outlet. Conserving the slope can be difficult and a little time consuming because it is not an easy task, but it is necessary for your gray water system to work properly. This is another reason you might need a professional plumber for this portion of your gray water system creation.

Traps, vents, and cleanouts

Every sink in your home has a U-shaped pipe or pipe connector, about midway down the line leading into the wall or floor. This is called a P-trap and is designed to contain a bit of water. The purpose here is to block the fumes or waste that can back up from your septic system or sewer pipe. Because of this "holding power," these P-traps are also prone to clogs as solids get caught and build up over time.

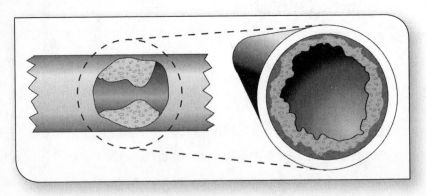

▶ Here, a plumber's snake is inserted into the cleanout.

Another essential plumbing element in every house is the vent pipes or vent stacks. These pipes are connected to the main plumbing lines and extend through the roof or out the side of the home. The vent stack allows built-up sewer gases to escape outside so they do not enter your home. This pipe additionally allows air to enter the plumbing system so pressure is equalized when water drains down the pipes. As the water is flowing downward, air is flowing in and "pushing" the water out. Without this equalization in pressure, a vacuum would form in the line each time the drain is used, and the water would back up into your toilet, sink, or tub.

Important for all homes, but especially with a gray water system, is an easily accessible cleanout because there might be more debris than usual in your lines. A cleanout is a Y-shaped pipe with an additional small section sticking out where water flows through the pipe. This Y-section is capped with a removable cover that allows you or a plumber to run a snake into the line to clear out clogs. A snake is a long, spring-like cable twisted and pushed through a pipe to pull out whatever might be clogging the drain.

Designing appropriate access to your pipes

As you lay out your plumbing lines and install pipes, keep these features in mind. Design the system with adequate traps, appropriate venting, and cleanouts wherever possible. Keep in mind that most plumber's snakes can only effectively reach about 30-feet, so cleanouts must be spaced at intervals where the flatter areas still can be reached. You also can snake into the pipe from where your pipe leaves the house.

When designing the plumbing layout for your gray water collection, it is also important to place all diversions or access pipes downstream from existing vents or traps. This will ensure the air and water pressures are maintained so you do not create a vacuum in your system. This is another place to enlist the help of a plumber. Have him or her look over your design to make sure you are not creating lines that might later on cause problems. A plumber will recognize problems quickly and can offer suggestions for re-design.

Drawing up a preliminary plan

At this stage, sketch out a basic design for where you intend to place your system. Include where you want to put your gray water system, where the pipes will divert from water sources, such as your shower drain, and where they flow to a gray water cistern or pipes that lead directly to a drain field. Be as detailed as possible, especially in relation to where the pipes come together inside or where they will extend out from your building. Mark places where appliances and their drains are located in your house and where you need to divert the gray water. Be sure to mark the direction in which water is flowing on your property and where you intend for your gray water to flow.

Delineate, too, all the areas where you want to use the water because this will help you determine if you will even be able to get the gray water to where you want to use it — or if you need to plan on moving a few pipes or landscape beds for better access. When selecting an outside area to use your gray water, be wary of parts of your landscape that already receive high volume of runoff from rain. This runoff could flow through the gray water field and transport it to places you do not want it to go or to places of which an inspector might not approve. Areas below semi-permeable or impermeable surfaces are the places of the most runoff. These are places, such as just below drains from roofs, patios, concrete-covered areas, pavement, and driveways. *In the next chapter, you will learn more about rainwater and diverting runoff so it will not affect gray water areas. Additionally, the Assessment Worksheet in Appendix C offers some simple instructions on drawing up your plans on grid paper.*

At this point, the sketch will be simple and should give you an idea of your best options. This drawing should be enough to give to your SWCD or contractor to work with and will likely be enough for the initial permitting approvals. If you are working with a contractor or SWCD office, they can help you determine the best design for your space.

Calculating and using gray water output

Once you have evaluated where and what type of gray water sources you have, you can use this information to help determine how much gray water you are producing. This will assist you in determining the size of your gray water system. The average home without aerators or other low-flow fixtures creates 55 gallons of gray water per person per day. The equation for a four-person household would look like the following:

55 gallons × 4 people = 220 gallons of gray water per day

This is an average, and there will be days when you produce much less and other days when you will produce twice as much. With conservation practices and water-efficient fixtures and appliances, you can reduce your output to about 40 gallons of gray water per person per day.

If you intend to send your gray water into your yard for irrigation purposes or to filter it naturally through the soil, check the soil percolation, also called soil perc. *Tips on doing a simple perc test are shown in the sidebar.* This shows how quickly the soil absorbs water and is determined by the type of soil you have and the soil's clay content. The perc rate of your soil is important only if you intend to divert gray water into your yard. If you only plan to use gray water inside your home to flush your toilet or water your plants, soil perc is not important and you can skip this step.

If the perc rate is too slow, the water will sit on the surface in a puddle and can begin to get a noxious odor to it, and if it is too fast, the gray water will not have enough time to filter and can contaminate your groundwater. Soil with a high clay content will have a slower perc rate, while sandy or gravelly soil will have a faster perc rate.

Soil Percolation Test

This simple test can be performed to determine the soil's percolation rate. Be sure to test the area where you are planning to have gray water flow.

1. Use a garden trowel (small shovel), and dig a hole in the soil 6 to 12 inches deep.

2. Repeat this in all areas you wish to test.

3. For each hole, you will need a wooden stake, such as a tomato stake, to mark the depth of the hole. Use a ruler, and mark inches with a pen from the bottom of the stake upward.

4. Stick the marked wooden stake in each hole.

5. On a piece of a paper, make notes of the depth of each hole.

6. Leave the stake in the hole while you fill it with water. Note on your paper how far the water drops in a given number of minutes. For instance, the water might drop 2 inches in five minutes.

7. Repeat the procedure a few times until the rate is the same multiple, consecutive times. This is your percolation rate for that particular area.

8. Repeat steps six and seven for each hole.

9. Convert your percolation rates to rate per minute. For instance, if your drop was 3 inches in six minutes, you would convert it using the following equation:

Number of minutes ÷ Number of inches = Minutes per inch

Example: 6 ÷ 3 = 2 minutes per inch

If you find that it takes hours for the ground to absorb the water or if you pour the water and it immediately disappears, you have a problem with the ground's percolation rate. There are solutions to these problems, such as creating a gray water system that works with your soil. In fast-moving percolation, you can add mulch and plants over the soil and around plants in the area you wish for the gray water to soak. This slows the water flow down and allows the root systems of the plants to purify the water. In slow percolation, you can create what is called a constructed wetland. A constructed wetland does not speed the water flow; it adds plants, gravel, and other materials to purify the gray water and deter unhealthy microorganism growth or a mosquito habitat. It does this by creating a specialized ecosystem.

If you wish to create a gray water disposal area for purification purposes, use the following chart to determine how large an area you will need based on percolation rate and the amount of water you plan to dispose of daily. The gray water rate is the amount of water the particular area can handle per day. The area needed is the square footage of the gray water area needed per gallon per day.

Soil percolation rate (minutes per inch)	0-30	40-45	45-60	60-120
Gray water load rate (gallons per day per square foot)	2.5	1.5	1	.5
Area needed	.4 (sq ft/ gal/day)	.7 (sq ft/ gal/day)	1 (sq ft/ gal/day)	2 (sq ft/ gal/day)

Suppose your area's percolation rate is 41 minutes per inch. According to the chart, the load the ground can hold is 1.5 gallons of gray water per day. The final number is the important one because it tells you how much space you will need to have available for a gray water system. In this case, it is .7 square feet for each gallon of gray water. If you have a family of four and you are not using any system of water conservation, you will use the following formula to determine the amount of gray water produced daily on average:

55 gallons × 4 people = 220 gallons of gray water per day

Now multiply 220 gallons by the number of square feet you will need based upon your particular soil's percolation rate:

.7 square feet × 220 gallons = 154 square feet

This only applies if you want to dispose of your gray water. If you want to use the area for irrigation, you will use a different type of assessment to determine whether you have enough gray water for your plants or yard. Please note that all of these numbers are rough estimates. There are many different factors to consider, such as climate, rainy seasons, droughts, and components to the soil and environment, so it is more of an art than perfect science. These formulas are meant to give you an estimate to help you make some decisions before you choose which type of gray water system to use and where it should be located.

If you wish to use your gray water for irrigation purposes and want to make sure you have sufficient gray water to accomplish that task, estimate that you will you use a half gallon of water per week per square foot of garden or lawn. Different plants have different water needs, but this average can help you make decisions about how much gray water you will need for your irrigation purposes or how much square footage can be irrigated based on how much square footage of garden or lawn you have.

Suppose you have the following:

- Flower gardens: 200 square feet
- Lawn: 400 square feet
- Trees and bushes: 100 square feet

This equal 700 square feet that will need to be watered. Using the formula, you can determine that your garden or lawn will use the following amount of water:

½ gallon × 700 square feet = 350 gallons of water per week

If you are producing 55 gallons of gray water per day, you can determine that you are producing the following amount of gray water per week:

55 gallons × 7 days = 385 gallons of gray water per week

According to this equation, if you were to use all your gray water to irrigate your lawn, trees, and gardens, you would have plenty of water. Again, these figures are an estimate, but at least in the example, using gray water for irrigation water would be feasible.

Keep in mind that you might have long periods of time when you will not need to water your yard or plantings, such as during the non-growing season or during stretches of rain. These factors need to be taken into account when designing your system because your home will continuously produce gray water, whether you need it for irrigation or not. Also, if you are planning to use a system with a pipe coming directly out of your house, make sure it extends far enough out from the foundation so water will not pool up against the walls of your basement. This can cause major damage to your foundation and water damage to the inside walls.

TIP! Make sure the pipe coming out of your house also diffuses the water enough so it does not come out in a powerful stream. This would eventually erode the area near the end of the pipe, causing the water to pool and flow to areas you might not want it to go.

You will also want to assess other factors, such as a tendency to flood in certain areas due to the slope of and permeability of the ground, to determine whether you can use gray water for irrigation or whether collect gray water to dispose of it naturally. It might not be practical to use the water for irrigation if the water cannot drain into the soil properly or will create flooding problems on your property.

Once you have completed your assessment and laid out your plan, it is time to get any required permits. Depending on your particular project, you might need a inspector's visit at various stages, including excavation, foundation laying, wiring, and plumbing. Your permit will outline these steps. Be sure to follow them to the letter because an inspector can make you remove finished work to inspect completed stages. In addition, start working with your utility companies to identify underground lines that you will need to be aware of before digging.

Construction and Maintenance Concerns

Most gray water reclamation projects will require heavy-duty work, especially in plumbing, excavation, and groundwork. When done incorrectly, these projects can fail, and you could end up with a basement filled with gray water or a large stagnant pool in your backyard. Even worse, you will be out the money you spent to construct it, and you will have to spend even more to clean up the mess. Before you even begin construction, be honest with yourself about your ability to complete the construction of a system on your own. Do you have the construction skills to build a gray water system? Do you have plumbing skills, or will you hire people to help you? Do you have the necessary skills to complete the work so it will pass an inspection? Do you have the time to complete this project quickly once you have begun? After all, once you tap into your gray water, you do not want to wait to finish.

Another important consideration is whether you will be able to maintain the system once it is built. There are not difficult or time-consuming post-construction maintenance issues, but using a gray water system will require a change in your normal way of life. It is important to make sure all family members are willing to learn how to work with the new methods of water usage. You do not want to invest time and money into a system you will eventually abandon because it is too much of a lifestyle commitment.

You might need to have your site surveyed and inspected before you begin any type of construction. During this time, begin to make a list of supplies you will need for your gray water system. You might also need to buy or rent tools and equipment, including table saws, backhoes, or cement mixers. Plan what you will need in advance, and make the necessary arrangements. If you are doing the work yourself, much of the needed equipment can be rented by the day or by the hour from your local hardware or home improvement store.

You might determine you cannot do all the work alone and that you will need assistance from other professionals, such as plumbers, to assist you. If you are calling in help, you need to decide who will function as the contractor. This person will coordinate all workers, supplies, and schedules so everything works together. Homeowners can take on this job as long as you are familiar with the design and have the time to devote to the project. As with any construction project, you can also hire a contractor to handle these coordination tasks.

The possibilities for gray water systems range from simple to complex. If you feel adventurous and want to tackle more complicated designs, check out Art Ludwig's *Create an Oasis with Greywater*. This book offers more information on establishing a gray water collection system for your entire yard.

5

Installing a Rainwater Harvesting System

▶ A water cistern was installed to collect rainwater that runs off the roof.

The principles and techniques used for rainwater harvesting are similar to those used in gray water harvesting. Your basic goal is to catch, collect, and use the rainwater that falls onto your property. Whether it is running through your roof gutters into a rain barrel or funneling through a series of grassy berms into a pond, this water is usable for your household. Rainwater touches many surfaces before it gets to your tank. For this reason, it is not potable water unless you filter it and purify. *Methods for purification and testing are covered in Chapter 6.*

There are a few differences, however, between rainwater and gray water. Rainwater collection is not as reliable as gray water because the source of water (rain) is intermittent and unpredictable. You can assume your area will get the average rainfall over a season, but you do not know when or how much rain will fall on any given day. When it does rain, the volume of rain might exceed your storage capacity. Also, during large rainstorms, this rapid influx of water will temporarily stir up any water already being held in your tanks. This will soil your stored water, and you will have to wait for the debris to settle back down to the bottom of your tank. For this reason, many people choose to keep separate barrels: one for previously collected or treated rainwater and one for newly fallen rain. If you are only using rainwater for landscaping or emergency storage, you will have time to wait until the next required watering.

In the last chapter, you learned the basic concepts and techniques behind creating gray water systems. Many of the issues related to rainwater harvesting are the same as those covered in gray water systems. This is especially true with permitting, regulations, foundations, and tank collection. These are important areas that must be addressed, so please refer back to the last chapter for specifics on these issues. This chapter will cover issues unique to rainwater harvesting.

Goals and Considerations of a Rainwater Harvesting System

As with gray water systems, before you fully commit to rainwater harvesting, it is important to determine your goals and then assess the issues you will face with this change to your lifestyle. Think clearly about what you expect from your water harvesting system. Note what you hope to gain and how much work you are willing to put into building and maintaining a system. Walk around your property, and envision how your new system will fit in — be especially aware of access and usage issues. The considerations are similar to a gray water system. *Appendix C includes an assessment worksheet to help you develop a plan, including the size and scope of your project and narrowing down the costs associated with your plan.* To fill out the worksheet, start by considering the following issues as they relate to rainwater harvesting:

- **Do you have enough space to set up your rainwater harvesting system?** Most rainwater harvesting systems require a rain barrel or an even larger cistern to collect rainwater. Do you have enough space to place rain barrels around your home, and will they be accessible to your yard or garden space? If the collection barrels are far from the areas where the water will be used, you will need a way to transport the water, which will mean you might have to buy a water pump, extra hoses, or buckets.

- **How will rainwater harvesting fit into the existing landscape around your home?** Will you or the neighbors not want to look at the collection barrels? Will you need to add foliage or screening to cover the system?

- **Are you planning on using mechanical additions in your system?** Do you intend to pour water out into buckets, or will you use a water pump or water filtering system?

- **Will your rainwater catchment system provide water for household use, or will it be used outside only?** How efficient will it be, and how many different areas will it provide water to in or around your home? Will your rainwater be used to irrigate your lawn and gardens? Will it be used to fill a pond or other similar structure?

Will it be used to flush the toilets or for showers? Do you intend to have a vegetable garden with a drip irrigation system? How much digging and other work will it take to get the water from the cistern to your garden? The more complicated the system, the more the project will cost.

⊕ **Is your goal to use your rainwater for drinking?** In some areas, this might not be legal, or it might mean you have to install a filtering system, change your roofing material, or get additional permits.

⊕ **Are you afraid of heights?** Installing a rainwater harvesting system might require you to climb on top of your home to install gutters. If you are uncomfortable with high conditions and are not used to working from a ladder or the top of your roof, you might need to hire someone to help you.

⊕ **Have you fully researched the laws about rainwater harvesting in your area?** There are laws and regulations you must follow, and in some states and areas, it is illegal to install some forms of rainwater harvesting systems. It might be possible to find a form of rainwater harvesting not covered by local or state laws, so do not give up on your first try. Review all the different types of rainwater harvesting systems, and be creative.

⊕ **Do you have the right climate for a rainwater harvesting system?** Sometimes an area's climate might not be practical for rainwater harvesting, such as areas that do not receive rain or areas where the ground stays frozen much of the year. However, there are systems that can be turned on and off by season and require more work in the fall and spring. An example of this type of system is a rain barrel positioned underneath a downspout. There are dew collection systems you can consider in the more drought-prone areas of the country. These systems collect water droplets from condensation that occurs when air cools overnight.

⊕ **What is your budget for a rainwater harvesting system?** As with gray water systems, putting in a rainwater harvesting system might be initially expensive. Before you begin pulling down gutters and buying rain barrels, research the costs associated with building the system and set a budget for your project. Also determine the cost of permits where they are required before you begin working on your rainwater harvesting system.

A system of this type is a big decision, as is a gray water system. Make sure you have thought about all the advantages and disadvantages before you begin buying or planning any rain-catching devices.

Legality of Rainwater Harvesting in your Area

Begin your search for legal information the same place you looked for information on gray water laws — the Internet. First, perform a search for any restrictions or limits on gray water usage in your area using Google, **www.google.com**, or another search engine. This search might give you all the information you need, or it might supply you with the contact information of the appropriate parties.

Rainwater harvesting has less need for regulation than gray water. For smaller systems, such as rain barrels attached to your rain gutter, you do not need a permit, and some municipalities encourage this sort of water usage. It is wise to check with your local regulatory boards before proceeding with a full-blown system. Ask the same questions you asked for gray water harvesting, and check with the same offices. Again, you might find helpful resources, assistance, and incentives from these experts. The primary issue you will likely face is concern about "diversion" of rainwater from neighbors or other sources dependent upon this water. For example, rain is a significant source for recharging ponds, streams, and wells, and during times of drought, these resources could dry up if the water is diverted away from its natural path. This would pose a problem for the people living downstream from you who depend on that water for their livestock or household use. In most areas, this is not an issue, but be sure to ask before you get too far into paying for or building a rainwater system.

If your research determines you will need permits and inspections, start this process early. You will need to present your preliminary plans to your zoning offices to secure the permit. Again, expect to provide more information because this is not a project most zoning offices are accustomed to handling. You might also be asked to come before the zoning board to address concerns. Throughout the construction process, you most likely will need several inspector visits, and the inspector can make you remove finished work if it was not properly inspected. Stay on top of these needs as you plan, and work to avoid later problems.

Tax or utility incentives for rainwater harvesting systems

As you begin to look into the laws concerning rainwater harvesting systems, you might also look into possible tax incentives or utility rebates available. A place to start is the website Harvest H2O, **www.harvesth2o.com**, which is a resource, no matter which state in which you reside. In addition, your local SWCD can provide you with the most up-to-date information regarding current programs available in your state. Call your water supplier, too, as they might have private incentive programs available. These programs are constantly changing, so be sure to check often as you work through the planning stages. Be sure of rebate amounts and incentives you will receive before you invest money. Keep in mind, too, that many of these programs are based on specific schedules and systems used, and they might even be first-come, first-served for funding.

Determining your property's watershed and rainwater potential

To begin designing your collection plan, you must first assess your property to determine your property's watershed, or the area where water enters, moves, and eventually stops on your property. Depending on the topography of your land, this watershed might begin far away from your property line or end abruptly at a gutter or stream. Consider where your highest point of influence is in your watershed. This could be at the top of a hill on your property or even the highest point on your rooftop. Determining the highest point can be accomplished using your eyes because a rooftop is higher than the ground below it. Another way to determine the water flow is to watch the direction water moves along the ground while it is raining. From this highest point, work your way down to the bottom and the area where water is accumulating in pools or soaking into the ground and no longer running off.

Knowing how water flows on your property is important because from this, you will know where to place your water harvesting system. You want the system collecting water higher on the property because it is easier to manage, as the water is flowing more slowly and with less force. As it moves downward and through the watershed, it will pick up speed and more velocity and make it harder to manage. Additionally, lower points on your property will have more water volume as it slows down and begins to accumulate. Also, by knowing the flow pattern, you will know which direction the water can naturally flow from your rain barrel or cistern to your lawn or garden bed.

The other thing to consider is where water on your property falls on impermeable surfaces and either runs off or begins to pool. If the ground cannot absorb it, the water will begin to flow down the slope of your lawn. This is referred to as runoff. The farther down the slope you are, the greater the amount of water that accumulates and the greater the speed of the water. The goal of rainwater harvesting is to turn this runoff into water that soaks into your lawn and garden or that can be stored.

A topographic map, which is a map that shows elevations on property, is a more scientific way of determining the elevation on your property. You can purchase topographic maps online or find them at your local library. These are organized by township or county, and with some research, you will be able to find your particular property on one of these maps. You might also be able to find this information at your SWCD office, and they might even do a topographical survey for you of the area you are planning on using. Circles on the map indicate elevations, and a number along the edge of the circles gives specific elevations. The closer the rings are, the steeper the bank. The center of the circles is the highest point, so water will naturally flow from this point toward the outer rings. Handheld GPS devices are another way to determine the elevation of your property. These devices use satellites to determine your location on the planet, and newer models can display your elevation.

The following image is a topographic map of a section of Utah. The numbers shown in the map are the actual elevations, which allow you to easily determine which way water would flow from higher numbers to lower numbers. The number zero is sea level.

Topographic map

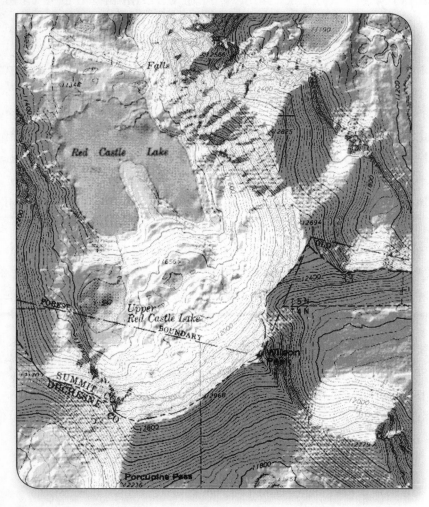

▸ Utah topographic map. Utah GIS Portal, **http://gis.utah.gov/sgid-image-server/
imageserver-new-shaded-relief-topo-layers-2**.

Determining rainfall amounts

The amount of water you will be dealing with, whether it is diverted for irrigation or for storage, is important. This amount will help you decide how much you want to collect, how big of a container you will need to collect it, and how much you can expect to put away for long-term storage or immediate use. You never know how much rain will fall in any given day, month, or year, but this will give you a baseline with which to work. The best guide is to begin with your average monthly and yearly rainfall amounts. You can find your annual rainfall averages on the Advanced Hydrologic Prediction Service website, at **http://water.weather.gov/precip**. This can help you estimate the amount of rainwater in inches that falls in your area. There are online calculators to help you convert these inches into gallons to see how much you are collecting. You can also ask at your local SWCD for help in coming up with an estimate of how much water you can expect to collect. *Appendix C also includes formulas for calculating and converting inches into gallons.*

Once you have determined the total rainfall for your property, divide this into the potential collections by specific surfaces or catchment areas. Use your sitemap to determine what the rain volume would be for the various catchment areas on your property, such as your home's roof, the driveway, a shed, or any other area that has a slow soil perc. *Refer to the soil percolation test earlier in this chapter for more information.* If the perc rate is too slow, the water will become runoff on your property.

If you are collecting rain from an impermeable surface, you can estimate how much rain you can expect to fall on your property during an average rainfall. Cisterns and rain barrels come in different sizes, and knowing how much water you are dealing with will help determine the size and number of barrels or cisterns to use. Keep in mind your earlier decisions about how much water you need to collect and store. There is no point in collecting 3,000 gallons of water if you only need 2,000 gallons. Also, be aware that collecting rainwater might divert water that naturally flows into areas that use it, such as trees, flowerbeds, or large gardens. Again, let nature work at its best, and do not redirect water that is already performing to its best potential.

Rainwater catchment types

From manmade to naturally occurring, rainwater catchment can occur in numerous forms. Some of these are simple and require little upfront cost or construction, such as a simple rain barrel. Others are complex and need extensive design, maintenance, and technical knowledge to use, such as earthen berms and holding ponds. Your system can combine whatever types work best for your property and eventual usage plan. The catchment types fall into these three categories:

Roof catchment type

 This is the most common type of system and directs rainwater that lands on the roof into a containment unit, such as a rain barrel. The water can then be stored, pumped, filtered, and used in many different ways. When considering this type of system, keep in mind that roofs that are galvanized, corrugated plastic, corrugated iron, or tiled work the best for this type of system. Flat roofs created from rolled roofing or cement do not work as well because they do not have a steep enough pitch, or slope, to allow the water to run off. There are different ways these systems can be set up on a roof and different options for containment units.

Ground catchments

These systems collect water that lands on the ground. The areas can be created from natural materials or manmade impermeable surfaces. Ground catchments can also be designed to siphon the water into a covered holding tank or could be just a barrel sitting on the ground to catch falling rainwater. People often use pre-existing paved surfaces, such as roads, runways, courtyards, or parking lots, as ground catchments because the rainwater will naturally flow off these impermeable surfaces. However, these surfaces can also contain pollution, such as oil, and the water might not be of the best quality. Ground catchments can be useful if you do not have a suitable roof. They do have the advantage of having a larger surface area for collecting rain, but on the other hand, these containment units often must be buried or stored underground.

Rock catchments

A rock catchment uses a naturally occurring rock outcropping from a rock wall. Normally, the water will run off the rock due to gravity. In a rock catchment, a wall is built on the lower part of the outcropping to prevent the water from flowing off its edge. This creates a rock pool. These types of catchments are used for communal water and are not used in homes. They are mentioned if you have a large flat rock surface from which you want to consider harvesting rainwater. The water from these types of catchment systems is directed through stone and cement gutters to a reservoir constructed using concrete or stone dams. These are large structures that hold enough water for an entire community.

Fog and dew collectors

These are simple units that allow water to condense on screens and then are collected at the bottom of the unit. These can be freestanding and mobile and can collect 1.5 liters of water per unit per night. Similar systems have been used for centuries to collect water in especially arid places in the world. They might not yield a huge amount of water, but the water is clean and can be used for drinking. If you are unsure about its purity, you can add a drop or two of bleach to a full container. Water from these collectors can also be used to water plants. It is possible to build your own dew or fog collector, and numerous plans are available online for this project. Visit **http://www.oas.org/dsd/**

publications/unit/oea59e/ch12.htm for thorough discussion and instruction for building and using a fog collector.

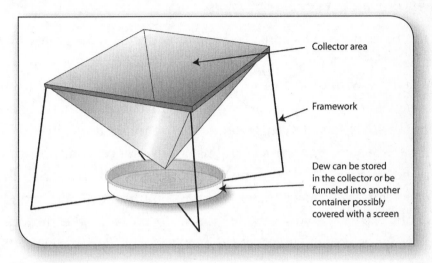

Collector area

Framework

Dew can be stored in the collector or be funneled into another container possibly covered with a screen

▶ Dew/fog collector

Working with a gray water system

If you are also using a gray water system, make sure any rainwater harvesting techniques do not interfere with the areas designed to collect or hold gray water. An example of this is when your rain barrel overflows during a heavy rain, and the water then washes through the area where your gray water is stored or used. This influx of rainwater could then carry the gray water into an area, such as your vegetable garden, where gray water would not be safe. Work through these usage considerations as your design your complete system. You cannot plan for every contingency, but thorough advanced planning will help you avoid some of the worst-case scenarios.

Designing a Rainwater Harvesting System

With these determinations, you can now decide which rainwater harvesting system you can use and how much water you will be dealing with on your property. When beginning your design, consider all these catchment options in conjunction with the unique attributes of your watershed. You might be surprised by all the ways you can harvest and use rainwater on your property. Try not to focus solely on one aspect, but rather consider all the ways you can collect water, improve infiltration, and reduce runoff. One way you can do this is to consider the multiple functions a particular structure can perform. For instance, the captured water in a rain barrel can be used to water your roses, while culverts divert water that can be directed to your vegetable garden. The efficiency of the entire system will increase if you can create multiple ways to use rainwater, rather than simply focusing on one possible use. In addition, the structures in your rain harvesting system can perform other non-harvesting functions, such as creating a garden space, creating a privacy fence, and providing shade. Spend the time, and think about the different ways water harvesting structures can enhance your property.

Keep in mind, as you are creating your water harvesting system, a primary goal is to capture and redirect rainwater so it can be reused more efficiently or spread out over a greater surface area, rather than being channeled into narrow runoff areas where it can cause erosion or not efficiently absorb. Another goal of rainwater harvesting is to gather it for storage for later use — and this could be far into the future. By creating runoff areas that allow for soil absorption, the rainwater will eventually make its way back into the aquifer. Rainwater harvesting is more than just catching water in a rain barrel. It can also entail deliberate planting to slow down the flow of water, the building of structures to redirect water and slow its progress, and thoughtful combinations of systems to maximize the water that comes into your property.

Use your site plan drawing to add where you will install rainwater harvesting structures and where you will need to divert water. Also note which way the water will flow over and through your property. You can use arrows to show the direction water will most naturally run. If you are using a gray water system, be sure to note this on your sitemap. You might not want to collect water for drinking below a gray water system. You can increase the efficiency of both systems by having them work together. If there is a strong, heavy runoff that crosses your gray water area, you might consider diverting the flow or collecting the water above the gray water area. You might have both systems working to infiltrate water in a particular area of your lawn. When it is not raining, your gray water system can continue to feed water to a particular area, and during the rainy season, you can divert rainwater for infiltration in that same area. Just be sure not to use any water gathered below a gray water area for anything except irrigation.

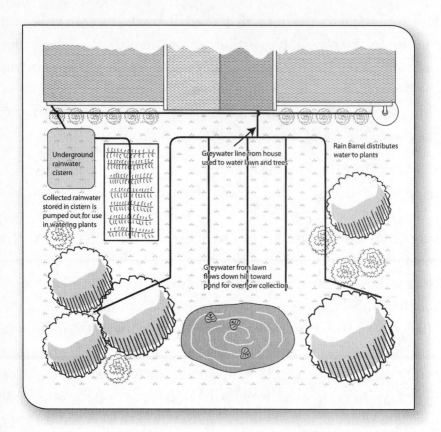

Underground rainwater cistern

Collected rainwater stored in cistern is pumped out for use in watering plants

Greywater line from house used to water lawn and trees

Rain Barrel distributes water to plants

Greywater from lawn flows down hill toward pond for overflow collection

▶ Sample yard diagram where rainwater and gray water systems are combined.

The following sections highlight important differences and requirements of small and large rainwater systems. Your project will be unique based upon the size, scope, and location of your rainwater harvesting system. This is an ongoing process, and once you have created your rain harvesting system, you might want to find ways to improve or add to it. Continue to re-evaluate your system and find ways to make it more efficient, cost effective, and beautiful.

Many people choose to start with a simple, small-scale rainwater harvesting system. You can always add more features later, and by starting with a simple design, it will be easier to harvest the rainwater and maintain the structures. Also, if you start small and your design does not work, you can quickly and cheaply make changes to the system. Creating a water harvesting system is a learning process. No two structures or plots of land are the same, so it might take a little experimentation to get it right. It is far better to fix a mistake on a small scale than on a large scale. You might also find that your dream of a large system is not feasible.

"Large" systems refer to those designs that use extensive earthwork to redirect rainwater, large watershed reshaping to hold the rainfall before it runs off, or large cisterns or holding tanks. Many of the design principles for rainwater harvesting are the same as gray water, especially when it involves the flow and diversion of water. Gravity is always the ruler of water movement. Harnessing this power with your rainwater system is crucial to efficient collection and usage of this harvested water. Refer back to the previous discussions regarding gravity-fed systems and fall and slope in plumbing. These same mechanisms affect how well your rain barrel will work.

CAUTION! Rainwater is not safe to drink unless it is well purified. *Chapter 6 covers purification methods and offers helpful resources for setting up your own purification system.*

Small systems

The simplest system to use is to buy a pre-made barrel with a spigot, or you can make your own rain barrel. *Appendix D, Plan No. 4 shows plans for making your own small rain barrel.* These barrels are placed at the bottom of your downspout so the water from your roof runs right into the barrel. If you do not have a rain gutter system on your home, you will have to install this also for a rain barrel to work. One key feature to an easy-to-use rain barrel is to use one with a built-in spigot. From this, you will be able to turn the water on and off just as with a regular tap, and you will be able to attach a hose so you can water areas far away from the rain barrel.

For rain barrels and small tank systems, the do-it-yourself method will work, and you can just buy what you need off the shelf. When planning for and setting up a small system, it is still important to think through how you will use the water. For example, if your garden is located at the back of your house, you do not want to locate your rain barrel off your front downspout. Arrange your system so the water is easily accessed once it is collected. Access into the inside of the barrel is important, too, as you will have to clean the barrel out. Many community education classes are now offered that cover rain barrels. Look in your local paper or ask at your library for more information.

A small system can also be mean using natural means to increase infiltration of rainwater into your soil. You do not have to collect the rainwater first in a container to put it to a more efficient use.

If, after a rainfall, you notice standing puddles or rivulets creating erosion in your yard, you might have a spot to install a natural "catch basin." This could be a small garden, bush, or tree in the area where water naturally flows but does not properly infiltrate the ground. One example of how to do this is to plant a tree and leave a wide area around the tree that you will cover with mulch. The rainwater will flow here, the mulch will capture it, and the rainwater will slowly filter back into the earth to feed your tree and aquifer. This mulch will increase filtration and decrease surface evaporation of water that might be normally sitting in a puddle. Grass and other natural ground covers can also work like mulch to hold the water and naturally filter it.

 TIP! If you are redirecting rainwater to a garden, make sure it does not enter this area with too strong of a current. A constricted flow of water will be forceful and could wash out your plants.

Large systems

For larger watershed collection, many people choose to install earthwork berms, culverts, and even dams to direct the water. These all should be undertaken with expert assistance. The biggest consideration is planning the overflow path and using that overflow as a resource in your watershed. By directing the water in a downward pattern through your watershed, it can be slowed down and used to water your lawn or fill a small pond. If you direct the water in a zigzag pattern, you will decrease the flow rate, reduce erosion, and allow for better infiltration.

? What is a Culvert? A culvert is a large pipe made out of metal, concrete, or other impermeable material used to channel water flow. It can range in size and is frequently used under roads, bridges, or in ditches to direct heavy rainfalls away from the structure and prevent erosion.

▶ Homeowners installed a culvert to redirect rainwater from a hill into their pond. The water is then used to irrigate their landscaping.

You can do this by laying culverts in angles rather than straight downhill. Be careful when using culverts or pipes because during a heavy rainfall, a culvert can act as a water cannon. Water will back up in a culvert, due to the sheer volume of water trying to go through a pipe with a small diameter. The pressure shoots the water fast and hard, and you can often find the worst erosion on a property just below a culvert. A natural drainage ditch can slow the water down and allow a wider diameter for water to flow through and reduce erosion. Many people plant tall grass or place large gravel at the openings to avoid erosion and slow the water down as it flows out of the pipe. Culverts are also a way to divert rainwater away from gray water areas and, therefore, prevent the gray water from becoming runoff.

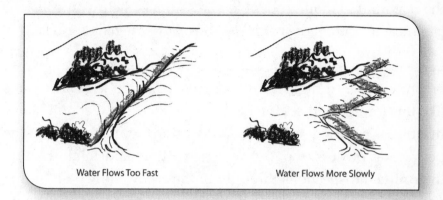

Water Flows Too Fast   Water Flows More Slowly

Large containment systems, such as tanks or cistern systems, must also be designed with a little more professional help. You are setting up a structure that will hold a large quantity of water, and if that structure fails, the resulting flood could be dangerous to people and homes near it.

In addition, others might currently be using the rainwater you are harvesting for their own watering needs. For instance, if rainfall on your property helps to feed a creek, which your downstream neighbor uses to water cattle, you might be taking away water on which he or she depends. This is not a deal-breaking issue but, as a neighbor, at least consider these issues if you are planning to collect rainwater. You will not have to abandon your plans for rainwater harvesting but might need a few adjustments so everyone is happy. Talk with property owners adjacent to your land, and consult with local officials before you start "tapping" into this supply.

Last, you only need to build a tank to hold just enough water for your use. If your watering needs only require 1,000 gallons of water, you do not need to build a 2,000-gallon capacity holding tank. A well-thought-out design will ensure your tank captures the maximum amount possible and will also help you consider all efficient methods of getting water out of the tank for use.

Large systems will require at least some sort of permit, approval, or professional design. Your local SWCD often provides plans like this for no charge. However, these offices can get quite busy during the building season, so start working with them early. The permitting process, too, can take some time and might require intermittent inspections throughout the excavation and building process. Inform yourself of these steps so you do not miss an important inspection. This book will cover projects that the experienced do-it-yourselfer can accomplish. Many of the steps in the projects, however, do require specific building knowledge, such as masonry, plumbing, or foundation construction. If you are unsure of any of these steps, hire a contractor or consult with an expert beforehand. Mistakes in these areas can lead to disastrous consequences, and, at the least, financial losses if the construction fails. Most permits also will require inspections of these trade-specific installations, and a licensed practitioner, such as a plumber or electrician, must complete the work. *Appendix A includes a list of resource books covering these aspects of construction.*

Selecting and installing a large tank

In most situations, you will be able to use a pre-made plastic, stainless steel, or fiberglass tank. Choose a tank meant to hold large quantities of liquid that can withstand the outward pressure this liquid will exert. Also, if you are planning to place your tank or cistern underground, be sure to select a material designed for this purpose. Not all materials can stand up to the pressures of soil pushing on the walls or the constantly moist soil against the outer walls. Some people with more time and money opt for building their holding tank out of cement block or even earthen structures. These latter options are appropriate for storage of water but require specific expertise.

The design and construction needs for these structures are complex. If you do not know how to handle these materials, it is best to consult with or hire a contractor to help you complete the project. If you are working with your local SWCD office, they can help you source materials and find contractors to build your tank. Any tank and its installation will have an expensive upfront cost, but it will be even more costly if your tank fails and you have to dig it out and start over.

Because of this large volume of water, your holding tank will be heavy and must be built on a foundation. *This topic was covered in Chapter 3, so refer back to that section for tips on building your foundation.* To set up your tank on the finished foundation, follow the recommendations of the manufacturer. If you live in an area where the ground freezes, you will have to install the tank below the frost line. Ask for assistance from your contractor if you are unsure of this step. Additionally, larger projects will require assistance from plumbers and, possibly, electricians (if you are using a pumping system). Again it is recommended to call in an expert to protect your investment and make sure everything is set up correctly. Larger projects, too, will require inspections at various stages of the project; these stages will be outlined in your permitting materials. After everything is hooked up, be careful as you fill in the hole and cover the tank with soil. Replace the backfill with care so you do not dent or puncture your tank. Make sure to sufficiently mark the area where you tank is located underground so you do not accidentally drive over it. Large equipment can easily crush a tank.

Keeping the Water Clean

When collecting rainwater, it is important to keep it clean from the beginning and as you store it up. By taking preventative measures right away, you will have less purification needs later on. The biggest issue to consider is stopping the influx of debris or pollutants the rainwater picks up as it flows into your barrel. This applies to the water collected off hard surfaces, such as your roof and driveways, or water that runs through an extensive watershed area. Common sense will guide you here as you set up and maintain your system. Keep in mind, it is easier to stop the debris before it gets into your barrel than to clean it out later.

Consider the path water will follow to your tank and take precautions to keep this route clean. Your specific system will require its own special care, but here are upkeep and preventative solutions common to many watershed areas:

- Install inexpensive gutter screens on your gutters so they do not fill up with leaves and dirt that will wash through your downspouts to your waiting barrel.

- Keep a screen on the end of the pipe leading into your rain barrel for one final catch of debris. Be sure, though, that this screen allows enough water to get through without backing up the water into the downspout or pipe. Locate this screen in an easily accessed area and clean it frequently so it does not get clogged.

- Keep your driveway swept if it looks like a rainstorm is coming, and avoid placing a collection line leading from a driveway stained with oil.

- For grass-fed drainage areas, avoid using fertilizer or weed killer in this grass because the water will pick up these chemicals and transport them into your rain barrel.

- For longer grassy or rocky watersheds, keep watch for dead animals, animal waste, or other toxic debris where the water flows.

- For large, open holding tanks, install a screen above the tank so frogs, insects, birds, and other animals looking for water are deterred from entering your water.

The materials used for roofing will also affect the quality of water as it comes off your roof. If you are planning on re-roofing or building from new, choose materials that will not absorb soil and bird feces or shed particles as the water flows across it. Fiberglass-based asphalt shingles used on most homes are porous and absorb more contaminants than smooth metal or tile roofs. The rainwater will then pick up these materials and flow into your barrel. Other shingling material, such as wooden shingles, zinc roof caps, or rolled roofing, are often treated with chemicals to waterproof them. These chemicals will leach into the water, can affect the quality of your collected water, and are difficult to remove if you are planning to purify your rainwater for drinking.

First-Flush Devices

One solution to a less-than-desirable roofing material is to design a roof-cleaning diverter or first-flush device into your rainwater collection plan. Rainwater will pick up the majority of contaminants as it first falls and runs across your roof. These devices redirect this first flow of rainwater away from your collection tank with the goal of carrying off the debris and contaminants and flushing them out of the system. After this initial diversion, which can range from 10 gallons to 60 gallons, the cleaner water shedding off your roof will flow into your holding tank. There are hundreds of diverter and first-flush designs available on the market. Prices range from $50 up to $1,000, and the device you select depends on your specific needs. Research those available by searching online for "roof cleaning diverters," or ask at the store where you bought your rain barrel for more information. Installation is quite simple, and most diverters are placed right into the pipeline or downspout leading into your holding tank.

Once the water makes it into your tank, it is important to keep it as clean as possible. As shown in the project plans, most rain barrels need a cover or at least a fine mesh screen. Larger systems are difficult to cover, so it is necessary to keep the surface skimmed of floating debris and occasionally remove any built-up sludge at the bottom on the container. Additionally, you need to keep bacteria, mold, and algae from growing in your stored water. Algae is not only unsafe to drink, but it can give off a bad smell and can even clog up the exit pipes coming out of your tank. The biggest contributing factor to algae growth is sunlight, which can get through white or light-colored containers. Choose black or opaque containers, or install a shade over your open-air tanks.

CAUTION! With proper purification and filtration, you could drink rainwater in an emergency situation. However, it is not immediately suitable, and even though it looks clean and does not smell or taste bad, it must be cleansed before drinking. Rainwater might even need to be purified before using it for livestock. *Filtration and purification processes for rainwater and other stored water is discussed in Chapter 6.*

Open, standing water is also an invitation for mosquitoes. Be sure to cover your barrel with a lid whenever possible if you are not going to use the water for a few days. Some mosquitoes can mature from egg to adult in four days and can carry disease to you and your pets. Larvae will not mature in moving water, so the most organic way to treat for mosquitoes is to use a fountain, waterfall, or aerator in your tank. You can use mosquito pellets sold at most home improvement stores. The pellets kill the larvae before they mature but are a pesticide, and water treated with these products will not be potable or organic. It can be used for irrigation, but additional clean water or rainwater should be added so the pesticide concentration is diluted. Without dilution, the amount of pesticides present could be harmful to your plant's foliage or flowers.

CASE STUDY: OVERVIEW OF A COUNTY RAINWATER PROJECT

Katherine Peragine
Catawba County Soil and Water
Conservation District
Environmental Educator
Newton, North Carolina

A new, cost-sharing assistance program was created in the state of North Carolina in 2007 to provide opportunities for Soil and Water Conservation Districts to partner with landowners in urban settings to address the issues of runoff, water resources, and erosion from development. The Catawba County Soil and Water Conservation District chose to use allocated funding to address a water-quantity need by installing large rain barrels and cisterns on public buildings throughout the county. Using retail examples as models, each water catchment system was tailored to the unique needs of the facility. Each example uses cubic feet of rooftop to direct rainfall into a collection unit. In one example, an Agricultural Resources Center uses rain captured in a 500-gallon rain barrel for filling Forest Service Pump Trucks and watering education landscaping and farming plots. Another uses underground cisterns to catch runoff then used to water school soccer fields. Another example uses the tin roofs of vehicle carports to catch water used for watering an entire county's landscaping efforts.

The following photos outline the process used in developing these systems and the installation process itself. In all examples, the process began by partnering with public organizations and informing them about the possibilities available to them through the cost-share process. In every example, in-kind contributions of material and labor during the installation process matched the state funding. Individuals can benefit from the photos and descriptions of this trial-and-error process, which has created useful and efficient rainwater harvesting systems on a large scale, and use them for personal home or business use.

Sizing the Cistern

▶ Satellite photo of the Agricultural Resources building in Newton, North Carolina (Catawba County Soil & Water Conservation District)

Due to the sidewalk that surrounds the building, the only place to install a cistern and not impede foot traffic was in an herb bed next to the fire escape. Then, we had to decide how many downspouts we could realistically catch.

Diagram of Building and Property

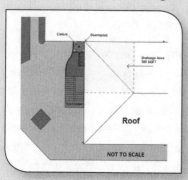

▶ In this diagram is the proposed placement of the cistern that is being placed under an existing downspout. This type of site map helps determine placement before any construction begins. (Catawba County Soil & Water Conservation District)

We calculated that the downspout we were going to intercept had an approximate 550-square-foot watershed. The North Carolina Community Conservation Assistance Program manual, which is the program that provides grants for rainwater systems for businesses and individuals, calls for 1 gallon of storage for every square foot of watershed, so our plan called for a 550-gallon cistern. Next, we had to find a cistern that would hold the amount of water we would be receiving and still fit in the area that we planned to install it.

Area before Placement of Cistern

▶ Downspout and bed before cistern installation. (Catawba County Soil & Water Conservation District)

The area we chose to place the rainwater harvesting system was agreed on because it already had an existing downspout, and there was enough room to place a large cistern. If there was not enough space, a different placement would have been considered.

Top View of Cistern Area

▶ The cistern site from the fire escape. (Catawba County Soil & Water Conservation District)

This gives an alternative view of where the cistern was placed. In the corner is the downspout that will be used for the rainwater cistern, and underneath is a flowerbed that would provide an area for the cistern. This place was measured out to make sure the foundation would be large enough for the cistern we were considering and also high enough.

Concrete Foundation

▶ Concrete foundation that will hold the cistern. (Catawba County Soil & Water Conservation District)

We poured a concrete foundation for the base of the cistern rather than using stone because part of the cistern would be sitting on the brick curb. The reason that concrete foundation was decided on was to level the cistern. Also, when the cistern is full of water, it will become very heavy and begin to sink into soil, so a firm foundation would prevent this settling. If the cistern settles, it can break loose of the downspout or not operate properly. We chose a 550-gallon cistern. The cistern is about 67 inches in diameter by 44 inches high. Empty, the cistern weighs about 100 pounds. Water weighs about 8.5 pounds per gallon, so when the cistern is full, it weighs about 4,775 pounds. This is why a solid foundation is important. If the tank were to break free of the building or topple over, the damage to property could be great, and the risk it could impose to human life.

Water Diverter

▶ The diverter ordered from Aquabarrel at a cost of $57. (Catawba County Soil & Water Conservation District)

We installed a diverter that would allow us to bypass the cistern if necessary. The diverter is the small piece of metal at the top of the downspout. When the cistern is full of rainwater, there is no longer need for the water to be diverted there. A diverter allows the flow to be directed to the ground when the rain barrel is full.

Water Diverter

▶ Pieces of the diverter before it is installed between the downspout and the cistern. (Catawba County Soil & Water Conservation District)

This diverter allows water to divert down the original downspout even if the cistern is not full. On the top piece, there is a level that sends the water down the water spout or toward filling the cistern. We use a long pole to move the switch when needed. The downspout had to be cut to the proper length to add the pieces of the diverter between it and the cistern. This can be done with a table saw or hacksaw. Each of the pieces should fit neatly in the next for a tight fit. No adhesive was needed.

Overflow on Cistern

▶ Top view of the cistern as it is placed on the foundation and secured to the wall. (Catawba County Soil & Water Conservation District)

We installed a bulkhead fitting to allow the overflow to return into the original drain. When the cistern is full, the excess rainwater needs to be able to escape, or it could damage the cistern and downspout. There is a pipe that extends near the top of the cistern and attaches to the downspout. When the water reaches the level of the pipe, it will begin to overflow back into the downspout.

Sight Gauge of Water Level in Cistern

▶ Site Gauge attached to the cistern. (Catawba County Soil & Water Conservation District)

We also installed a sight gauge to allow us see how much water we had accumulated. The gauge is attached to the side of the cistern. Water rises and falls in the tube according to how much water there is in the cistern. This moves the yellow ball up and down and can give an observer an idea how much water has been collected. This is important because the cistern is large and opaque, and so it is almost impossible to see inside or pick up.

Complete Cistern

▶ Completed cistern — set, plumbed, and anchored to the wall. (Catawba County Soil & Water Conservation District)

The cistern is now complete. All the pipes are attached to the cistern and the downspout is connected (also known as plumbed). A bolt was drilled to the wall, and a strap that surrounds the circumference of the cistern was attached. This keeps the cistern from moving or tipping. The site gauge is attached at the top and bottom side of the cistern with a long, thin plastic tube so the water level can be observed.

Signage on Cistern

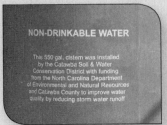

▶ The sign added to the side of the cistern to explain its purpose. (Catawba County Soil & Water Conservation District)

We included signage on the cistern to inform the public about who installed it, where the funds came from, and that the water is not drinkable.

Electric Pump for the Cistern

▶ Pump added to the hose to increase water pressure. (Catawba County Soil & Water Conservation District)

There is not enough water pressure from the cistern to be useful for watering the lawns, so we added an electrical pump. This allows us to pump the water through the hose to water lawns and gardens and even to wash the county cars.

Lawn Watered using Cistern

▶ Harvested rainwater used to water lawn and garden. (Catawba County Soil & Water Conservation District)

Water is pumped from the cistern to a water hose, which is attached a sprinkler head. With the electric pump, there is sufficient water pressure to completely irrigate the lawns and gardens around the building without the need of city water.

Water Quality

Having a water storage and collection plan in place is a start, but there is more to it than just gathering and storing the water. The quality of the water needs to be protected, especially if you are beginning with high-quality, pure drinking water. If the water's quality is compromised, you can become ill and even contract life-threatening diseases. You will also need to spend additional time and supplies to regain the water quality. In an emergency, or even for daily household use, spending extra time retreating water that has lost quality is time that could be better spent doing something else.

Water can be contaminated in numerous ways. Airborne pollution, contaminated runoff, and chemicals leaching from your storage containers can cause problems with your stored water. Even something as simple as a tiny bit of bacteria can multiply tenfold if a filled bottle is stored improperly. Collecting and storing all this water has taken you time and effort, and taking precautions from the start will ensure stored water retains its initial high quality. The following sections will offer ideas on how to protect the water you have stored, how to test and maintain it, and how to purify water you have collected.

The range of testing and maintenance options vary widely depending on how, where, and what type of water you are storing. You will need to test your water every few months or change the storage containers, and you will only need to check it as needed. As you select storage methods, keep notes on the testing methods required to maintain quality. Record on your monthly calendar when these tasks need to be accomplished so you do not forget. Consider keeping a small water storage notebook to track your stockpile and maintenance needs. Each time you make an adjustment to the system, empty a container, or test the water, write it down in your notebook. As discussed previously, try to rotate any emergency-stored water every six months and make sure to sanitize the containers before refilling.

Where is Your Water Being Used?

The level of water purification or treatment you need depends on how you will use your water. Water used for landscaping will require no purification, water destined for appliances, such as washing machines, might need filtering to remove heavy metals, and drinking water will require the highest level of purification. For instance, some metals and chemicals will cause buildup on faucets and in water lines while others can affect the workings of your washing machine and dishwasher. However, the bacteria present in these waters do not need to be removed for plain household cleaning, but they must be cleansed out for drinking water purposes. Knowing how you are going to use the water and what chemicals or metals are present in your water will help you determine the level and types of treatment needed.

CAUTION! Drinking water is not just the water you drink from a glass. It includes water used for showering, brushing your teeth and cooking. If there is any chance the water will be ingested, it must be purified. Do not forget your refrigerator in all this planning. If you have a fridge with an icemaker or ice and water on the door, make sure the fridge is filtering incoming water, or install a filter yourself if the machine does not have one.

If you are planning to use rainwater for your entire household, you can develop a series of purification plans to cleanse water for each use. With a thoughtful plan, you will not spend extra time and money over-purifying your water. For example, you might want to use a UV filter for your household-use water and add a carbon filter for any water you want to drink. *Further sections will highlight the available purification methods, and manufacturer resources are listed in the Appendix.*

Testing your Water

The type of water testing kit you need depends on how you are planning to use your stored water. Tests kits cover a variety of specific parameters, such as testing for pH levels, indicating the presence of heavy metals or nitrate, and checking for bacteria. If you are planning to use your water for drinking, choose the kit to test for bacteria. Even a small amount of bacteria in your water supply can make you and your family ill. If you are only going to use your stored water for your landscape, then you just need a kit that tests for nitrates or pH levels.

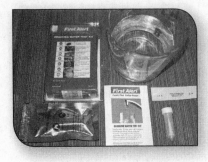

These test kits can be purchased from numerous online sources or at most large home improvement centers, and they are frequently sold through your local county extension or SWCD office. The test kit should have thorough and easy-to-follow instructions and will give you an automatic reading, such as a litmus strip that turns a certain color. Follow the directions carefully and avoid contaminating the water after you have taken the sample, such as with a dirty container or hands. For more extensive testing, you might have to send the sample into a lab for a panel test.

Water originally collected from your household tap (if your local water district provides it) that was stored in sanitized containers should retain drinking water quality standards without a problem. The same goes for unopened containers of water purchased for storage. The water might need aeration, or exposure to air, if it is flat, but the quality will remain intact, and do not worry about testing this water. However, if when you open the container you notice an off-color or bad odor, it would be safest to test the water before drinking. This is why it is an idea to store a few test kits with your emergency water supply so you will be able to test the water if needed. Make sure to check the expiration date on your test kits and replace as needed.

If the test kit reveals a contamination, you will have to treat your water or replace it. If time allows, you could use this water on your landscape, refill the sanitized containers, and start over. To restore quality for drinking and eliminate contaminants, you will have to boil it, treat the water with bleach, or use a purification system. *These methods are covered further in the following sections.* Some contaminants, such as heavy metals or chemicals, will not be removed by boiling or bleach — these require additional purification methods. If you filled the containers from your household tap and still found contamination, you will need to work with a professional to eliminate the source of pollution. Your local county extension office can help you find a solution to this problem.

Rainwater should always be tested before storage if it will be used for more than landscape watering. The contaminants present in rainwater can range from pollution to debris picked up off your roof or from your gutters. Even if you are planning to purify the water, you will still need to test it before purification to know what contaminants with which you are dealing. Additionally, rainwater intended for purification will always have to be filtered in some way to remove tiny pieces of debris. Even if you are using a purifying technique, you might consider testing your water after purification to make sure your system is working.

Before you go out and buy separate test kits for drinking water and rainwater, do research online or at your SWCD office to determine for what you will be testing. You might be able to use the same test kit for both purposes. Most test kits are a one-time use item, so be sure to keep enough test kits on hand to adequately test your water for the duration of the emergency.

? Always test water from a new well

Many disease-causing organisms and elements can be present in well water. It is important to have the water tested when you are digging a new well. One internationally known example of the hazards of well water happened in Bangladesh in the 1980s. A widespread effort was undertaken to dig wells for villagers so they would have convenient access to water. Unfortunately, the groundwater supply in Bangladesh contains high concentrations of naturally occurring arsenic. Villagers used the wells as their primary water source, and thousands suffered from arsenic poisoning in what the WHO called "the largest mass poisoning of a population in history."

Depending on where you live, elements, such as arsenic, fluoride, and lead, can be present in well water, springs, or creeks. In high enough concentrations, these elements can lead to serious diseases. Your well-digging company will recommend the testing procedure to screen for potential contaminants. Follow this regimen carefully, especially in your first year of well use.

Treating Collected Water

Rainwater, well water, or other non-tap water should be treated before storage so any bacteria or organisms in the water are killed. Just like in the emergency situations covered in Chapter 3, you can treat non-tap water by boiling it or treating it with chlorine. To review: The EPA recommends boiling water as the best way to kill bacteria. Water must be brought to a rolling boil and boiled for one minute, or three minutes at altitudes over 5,280 feet, to make the water safe for drinking. Once boiled, cover the container, and let the water cool down to room temperature before you replace it in the containers.

To treat the water with chlorine, use plain, household bleach in the amount of 1/8 teaspoon (16 drops) per gallon of water. Use only unopened, plain bleach, not bleach designed for colored laundry. After adding the bleach, stir thoroughly with a clean utensil, and let it sit, covered, for 30 minutes. After 30 minutes, the chlorine smell will still be there. If you cannot, retreat the water with the same amount, and allow it to sit for another 15 minutes. Always treat bleach as a potential poison. If the water retains too much of the chlorine smell, you can let it sit uncovered for several hours to dissipate, and transfer it from one container to another several times to add oxygen to the water. Just make sure you do not introduce bacteria during this "airing" process and container transfer. *See Chapter 8 for tips on improving the taste of treated water.*

CASE STUDY: ALTERNATIVE WATER TREATMENTS

Katie Cavert
Appalachian State University
Technology Department
Recycle at the Rock Volunteer
Coordinator
www.recycle.appstate.edu
cavertgk@appstate.edu

For people living in the developing world, accessing clean drinking water is an everyday struggle. Contaminated water carries pathogens, which, when ingested, cause diseases, such as diarrhea, cholera, typhoid, hepatitis, giardia, and worms. During my research into alternative water treatments in Third World countries, I have uncovered techniques that can be used in water storage and treatment to be sure that it is safe to drink.

These methods of water treatment are affordable and sustainable and have been found to work here and abroad. They include boiling, filtration, chlorination, and solar radiation. Boiling water is a common technique used to kill all microorganisms in contaminated water, but this can be an issue where fuel might be scarce. Bio-sand filters received high marks in all categories and performed the best in long-term, sustained use. However, this technology is the most expensive up front and costs between $25 and $100.

Ceramic filters also produce high-quality water and are moderately expensive, from $8 to $10 with an occasional replacement part costing between $4 and $5. The rate of sustained use is high until the filter breaks, which attributes 67 percent of the disuse rate. Chlorination treatments, including liquid solutions and tablets, treat a large amount of water, but unavailability of chemicals was a major reason for disuse, making this treatment solution less sustainable. Coagulation systems produce high-quality water but are time consuming, dependent on user hygiene, and expensive or difficult to find in the community, which resulted in user decline.

Solar water pasteurization technology, also called SODIS for solar disinfection, involves placing reusable bottles filled with water into full sun for about six hours to disinfect the water for drinking and the bottle itself as the drinking container. The water temperature needs to reach 65 degrees Celsius or 150 degrees Fahrenheit for 99 percent of pathogens to be killed (Solar Cookers International, 2009).

Using the SODIS method, instead of boiling to disinfect water, has several environmental and social benefits. It does not rely on natural resources as an energy source. SODIS is also one of the most affordable ways to disinfect water because the user only needs plastic or glass bottles. In addition, reusing the bottles will encourage recycling by the end users instead of the bottles being disposed of in landfills or incinerated with other trash, further mitigating air and ground pollution in developing countries.

However, SODIS treatment does not produce a high volume of water, and users found it difficult to tell whether or not the water was clean (Sobsey, et al., 2008). For the end user to know the water is clean, a temperature indicator is required. A commercially available soy-based Water Pasteurization Indicator (WAPI) has been developed for this purpose. The WAPI is a reusable, polycarbonate tube holding soy wax that melts when the water reaches sufficient temperatures. It is sold for $8 (Sun Oven, 2010).

Beeswax, a more easily found local resource in much of the developing world, is also being investigated as a visual temperature indicator (Cavert, 2010). Positive results could be disseminated in order to empower people in the developing world to acquire better health without much investment of money. It could provide economic opportunities to bee farmers and create local production markets and entrepreneurs in regions where SODIS technology is adopted.

Purification and Filtration Systems

Every drop of tap water you consume has gone through some type of filtering or purification method. Whether it is a large-scale water treatment facility or a simple filter under the sink, the water you drink has to be treated to make it safe to drink. The water pulled out of the ground has already gone through nature's own filtering system as it trickled down through the rocks and ground-cover above it. These methods, whether they are high-tech purifiers or low-tech sand filters, can be adapted for use in your own home. The version best suited to your needs depends on how you are planning to use the water, how much effort you want to put into the process, and how much you want to spend.

For example, if you want to be able to use rainwater, pond water, swimming pool water, or other options for drinking water in an emergency situation, you will need an extensive purification plan in place to handle this water. However, if you are only storing water for household cleaning or outdoor watering, it will not have to be put through much purification before you can use it. Develop your emergency plan and usage plan now by assessing the sources and quantities of drinking water you will need for each individual purpose. You can then collect the materials or machinery needed for proper purification. Learn each method thoroughly before disaster strikes so you can get your purification system running as quickly as possible. Also, make sure you follow the replacement and maintenance recommendations from the manufacturer so everything is up to date and in working order.

To make non-municipally sourced water potable, it must go through a multi-step process of purification. Most ready-to-buy treatment systems incorporate these steps, though companies use different systems for their designs. Furthermore, many of these more advanced methods of purification are only used in large, municipal applications and would be cost-prohibitive for the average homeowner. In a commercially made system, rainwater is harvested into a large tank. When a faucet in the house is turned on, water from the tank is pulled through a series of filters and into the house. Frequently, the system uses a combination of two charcoal filters and a UV filter to create the cleanest possible water. This system could also be combined with a large tank for holding rainwater outside and a smaller tank of filtered water inside, ready for the household to use.

The most common home-use methods are explained in further detail in the next sections. These treatment steps include:

- **Step 1: Filtration of the water.** Filtering options include a one-time filter or multi-cartridge approach. Technologies used for filters can be activated charcoal, reverse osmosis, nano-filtration, sand filters, or a combination of these.

- **Step 2: Disinfection of filtered water.** Methods for this include boiling or distilling, chemical treatments, such as bleach or tablets, ultraviolet lights, or ozonization.

- **Step 3: pH control.** This step is most commonly done with well water that might have a higher than desired nitrogen level.

Beyond these steps, large amounts of rainwater must also be screened for debris and allowed to sit to let remaining solids settle to the bottom of the tank. Most purification machines designed for home use are not made to filter out large pieces of debris or sludge. Keeping debris out before it gets to your collection tank is the best procedure. The simple systems cannot process large amounts of water at one time, so for household use, they need to be used in conjunction with two holding tanks — one for unfiltered water and one for the filtered water that is ready to be used.

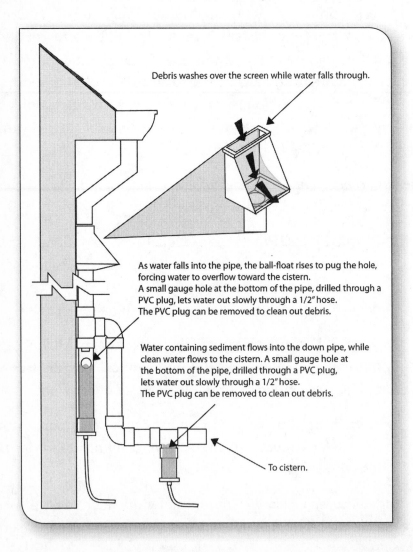

Debris washes over the screen while water falls through.

As water falls into the pipe, the ball-float rises to pug the hole, forcing water to overflow toward the cistern.
A small gauge hole at the bottom of the pipe, drilled through a PVC plug, lets water out slowly through a 1/2" hose.
The PVC plug can be removed to clean out debris.

Water containing sediment flows into the down pipe, while clean water flows to the cistern. A small gauge hole at the bottom of the pipe, drilled through a PVC plug, lets water out slowly through a 1/2" hose.
The PVC plug can be removed to clean out debris.

To cistern.

Caution! Just because water is clear or colorless, it is not necessarily safe to drink. Pathogens and microorganisms are invisible to the naked eye and are often odorless. Purifying your water is one of the most critical tasks you will do as a homeowner and during emergency situations. If done improperly, contaminants will not be adequately eliminated from your drinking water, and you will get sick from this water. Serious, life-threatening illness can result, such as giardia and cryptosporidium. These infections lead to vomiting and diarrhea, which can cause severe dehydration. In times of emergency, you might not be able to get the necessary health care needed to combat the resulting illnesses. Even during the best of times, water-borne illnesses can be debilitating and difficult to treat medically. In addition, the young, the old, or those with compromised immune systems will not be able to fight off the infections and could die.

Many of the more advanced filtration or purification systems can be quite expensive and require advanced technical knowledge to operate. The skills needed for construction, implementation, and maintenance of these systems can take years to learn and are not suited for a do-it-yourselfer. With assistance, though, you can adapt these systems to work for home water purification and filtering. Do not take chances with this step, and do not take on a project by yourself if you are not completely sure of what you are doing. Ask for help as you build and use your systems.

Information on where to find manufacturers for these systems is listed in Appendix A.

Screen or cloth filter

This method of filtering will not purify your water, but it will remove debris, such as bugs, leaves, or other materials that have entered your water tank. Large amounts of debris or dirt will eventually plug your hoses, so it is beneficial to remove it before you start using the water for irrigation. In addition, removing leaves and such early on will make further purification simpler and faster. If the material is left to sit in your tank, it will rot and decay.

If you find that debris is getting into your barrel, this do-it-yourself filter can solve that problem. Just make a frame, or use a large embroidery hoop to hold a piece of screen or T-shirt- type material taut. Suspend this filter in the top third of your rain barrel and well above where your exit spigot is located. This will trap the debris without clogging the hole where water must exit. You will have to clean out this filter occasionally so water can flow through it.

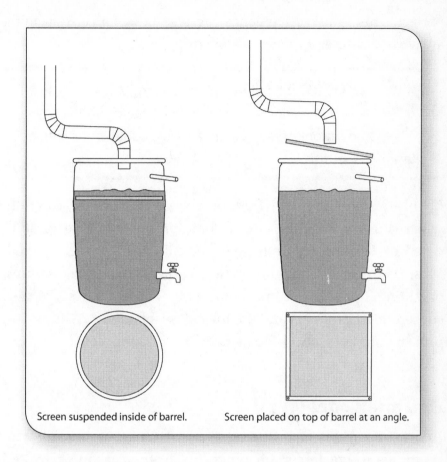

Screen suspended inside of barrel. Screen placed on top of barrel at an angle.

Chemical treatments: drops and tablets

Previous sections have discussed adding chlorine bleach to your water to disinfect it. You can also buy chlorine- or iodine-based pre-measured tablets or drops to disinfect your water. Readily available at outdoor-supply stores or online, these tablets are inexpensive and effective. Campers and back-country hikers have relied on this method for decades as a way to always have safe drinking water on hand when local resources are less than perfect. This same convenience can also be useful for stashing away with your emergency water supply.

 CAUTION! Some people are allergic to iodine, so be sure you are aware of this when stocking your kit. One indication of an iodine allergy would be a person allergic to shellfish.

If you are planning to use drops or tablets, read the label carefully when you purchase the bottle. Some products have expiration dates or become less effective once the package is opened. Most products also require stirring, certain water temperatures, or sitting times to be most effective. These tablets are also concentrated products and are toxic if mishandled. Wear gloves, protect your eyes, and keep away from children.

Activated carbon

Activated carbon is a special type of charcoal treated with chemicals to create certain properties in the carbon. Activated carbon absorbs harmful chemicals and particles from the water using its porous surface. These pores give the charcoal an enormous amount of surface area, which helps in adsorption. Adsorption is the process by which materials become attached through a chemical attraction. For water filtration, the charcoal adsorbs chemicals in the water and purifies the water. Activated charcoal will absorb many chemicals but will not remove them all from the water. Once the surface area is full, the charcoal will not continue to work.

Sand filters

This process involves allowing water to soak through layers of sand. A slow-sand filter can eliminate 90 percent or more of bacteria from water and process up to 180 gallons of water per day. A pressure-sand filter will work even faster because water is pumped through the sand at a higher rate. The sand works naturally to adsorb bacteria, pathogens, and small debris from water and cleans the water similar to an activated charcoal filter. The sand will eventually develop a gelatinous layer just below the surface and must be refreshed with new sand on a regular basis. This process of refreshing the sand layers can be tedious and problematic because the used sand must be disposed of, and new sand must be found to replace it. Organizations, such as the WHO, the United Nations, and the EPA, recognize sand filters as a viable water filtration option in developing nations and industrialized countries.

Constructing a sand filter is not a job for the do-it-yourselfer, though, as the materials and layering vary widely depending upon your region, the type of water flowing through the filter, and the final collection point of this water. A geological or agricultural engineer must design the filter, and a technician from your SWCD office supervises this step, along with the actual construction. This technician will teach you how to use the sand filter and offer maintenance tips to keep the filter operating smoothly.

Ultraviolet water filter

An ultraviolet, or UV, water filter uses ultraviolet light radiation to kill bacteria and microorganisms in water. A UV filter will kill mold, bacteria, and illness-causing oocysts (spores), such as giardia, but it does not eliminate chemicals and heavy metals in the water. Consequently, activated charcoal filters and UV filters are combined. For drinking water purposes, a UV filter and carbon filter together will kill most, if not all, illness-causing elements.

Faucet-mounted and pitcher purifiers

 These types of devices are sold at most large retailers, but there are many different versions available at reasonable prices. The faucet-mounted styles are installed right at the end of the tap or within the plumbing under the sink. These use a variety of methods to purify the water — again, read the label so the filter you are choosing will eliminate the contaminant you are most worried about. The drawback with faucet-mounted systems is they will only work if you have water coming to your faucet. During times of emergency, this is not possible, and it is difficult to use stored water with a faucet-mounted design.

The pitcher-type filter system is a simple design where the water is poured into a pitcher fitted with a self-contained filter through which the water flows. This filter slowly removes impurities as the water passes through. The methods used vary by manufacturer.

The main drawback with this is, the pitcher can only filter small quantities of water, which takes time to process the water. These pitchers are an addition to an emergency kit, though, as they are portable and convenient. Just make sure you have enough replacement filters on hand to get you through the duration of the emergency.

However, these types of filters are designed to work with already treated water, such as the tap water coming from your water treatment facility. Their main purpose is to eliminate any chemicals or heavy metals that might have gotten into the water as it traveled through your pipework, such as copper, mercury, or lead. Although this is important to your long-term health, these filters will not remove illness-causing bacteria that could be in rainwater. They are certainly useful as part of a water filtration system, but they will not work as the only purification method.

Pasteurization

This method uses high heat to kill any pathogens in the water, just as it is used for milk production. The water is pumped through a series of heated chambers and pipes, held for at least 15 seconds at high heat, and then pumped into a holding tank. This is a time-intensive, expensive process not suitable for the average homeowner. It also requires electricity to operate, extensive plumbing installation, and regular maintenance.

Purification machines

This is the most expensive method of purification but also the most reliable, portable, and convenient. Each brand has its own "secret" to purification, and these machines use combinations of the previous methods and can include everything from the methods mentioned previously to reverse osmosis to gravity-fed ceramic filters. The best purifiers on the market can remove virtually anything from the water, including metals, herbicides, pathogens, bacteria, microorganisms, and chemicals. Many models require electricity to operate, but there are versions available that work strictly on water pressure and gravity. These devices are also easy to operate, install, and only require plugging the device into an electrical outlet and hooking up your water supply. Some are even battery-operated and can be filled by hand. These ultra-portable systems are the top choice for campers and hikers, and international service organizations, such as the Red Cross, Unicef, and other groups operating in areas without potable water, use them.

Because of this variety, each purifier on the market has its own distinct advantage. Some offer faster filtration, some offer quieter operation, and some are more portable and space saving. Each company also offers a wide range of sizes from miniature, hand-held bottles to large multi-gallon tanks. Technology is constantly evolving, so new methods of purification are still arising today. You will have to decide which feature fits best into your purification needs, budget, and workload. Keep in mind that the upfront cost can be high, but the day-to-day operational costs are low — much lower than the fee most cities charge for regular water usage. These machines also do not require maintenance, other than occasional cleaning or replacement of the filters. If you are installing a purification system to use in an emergency, do not choose a machine that needs electricity to operate. Select a hand-pump or battery-operated device for this purpose.

FUN FACT! The British Berkefeld system has been in operation since 1835, when Queen Victoria commissioned them to develop a water purification system for the royal palace. This gravity-fed ceramic filtering system is still used today and is considered the industry standard for many systems.

If you are planning to purchase a purification system, the best place to start your research is online. Search for the term "water purification" or "British Berkefeld systems," and you will get a variety of company websites to view. Rely on websites that end in .edu or .gov. Read carefully, and be sure to compare each device fairly on similar merits. Look at the bigger home improvement store sites, and find pricing on a wide selections of machines. Once you have settled on a few favorites, talk with your favorite plumber or others who use purification systems for opinions on how well these devices work. Look, too, at online or magazine product review sites for hints on what others think about this product. Avoid purchasing a used machine because you cannot be sure if it works properly — and purifying your water is not something with which you want to take chances.

 TIP! Wash your hands before you start working with purified water; otherwise you could reintroduce bacteria into the clean water.

Purifying Large Amounts of Water

Purifying water on a large scale is a complex and expensive undertaking. You will need an extra tank to store and protect the purified water. This larger capacity can bring the cost and installation of a large purification system up to a significant amount. For large systems, it is highly recommended to call in an expert. Follow the same research of methods and brands as suggested above. This will help you narrow down exactly what you are looking for.

From this research, you can begin working with a water specialist, such as a plumber, water quality expert, or contractor from the company you have chosen. Larger systems will qualify for help through your local SWCD. Contact them first if you think a large system is right for you. With their help, you might be able to get a full design, usage and maintenance assistance, and possibly have the expenses paid through a federal or state grant. At the least, the people at the SWCD will have the best information on resources and products that will work in your region. They can also help you set up a preliminary buffering sand or grass strip to filter the rainwater before it goes to your more complicated purification system.

It is possible to combine numerous small purification systems — especially in times of emergency. This is not cost-effective and will take up your time for monitoring, maintenance, and usage. One advantage to a collection of small systems, though, is you will have backup during emergencies if one system fails. With a large, single-site purification system, you will have only one source of clean water during an emergency, and if that fails, you will not have access to water.

Keeping Your Storage
Systems Running Smoothly

As with any construction project or addition to your property, you must plan with maintenance in mind. Many of these systems, especially large or frequently used tanks, will have mechanical parts that wear out or break down. In addition, any system involving water also has the potential for leaks, the occasional need for drainage, and even the possibility of attracting rodents. Designing with all the issues in mind will make your maintenance and repair tasks easier to finish.

If you are working with a contractor or buying a pre-constructed tank, many of these access points will be designed from the start. However, your use might be considered "off-label," so make sure the inlets, outlets, drains, and service access fits with your water collection and usage plan. Be sure, too, that your contractor is aware of your entire plan so everything is installed properly for your intended use.

Each system has its own unique design, and this book cannot adequately cover every possible configuration available on the market. When researching your choices in storage tanks, though, it is important to consider the following issues related to maintaining and using a large storage tank. Think through your access points and service issues before you invest money and time.

Tank Inlets and Outlets

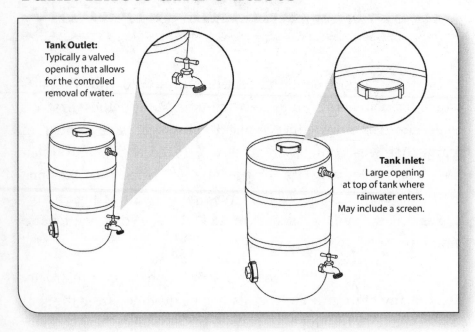

Tank Outlet:
Typically a valved opening that allows for the controlled removal of water.

Tank Inlet:
Large opening at top of tank where rainwater enters. May include a screen.

Water flows into your holding tank through an inlet and exits the tank from an outlet. In many tank designs, the inlet is near the top, and the outlet is near the bottom. This design allows you to use gravity for moving the water out. This also allows any sediment to settle on the bottom and not allow incoming water to stir it up. The outlet is located above this sediment layer, so it also does not pull the sediment out with the water.

Location of the inlet, outlet, and drain are specific to your holding system and to the area where you are gathering water. Placement of the inlet is at or near the top to allow the tank to fill completely and to help gravity move the water to the inlet. The outlet is placed near the bottom of the tank because an outlet placed too high will create a "dead storage" in the tank — an area of water below the outlet that will not be accessible once the water is drained down that far. These placements do not necessarily apply to storage tanks being filled by springs or a well or for people who will use pumps to remove the held water. For tanks that hold large amounts of water or for water-table wells, it is best to consult with an expert on where best to place these holes. Well installers, plumbers, and tank manufacturers have installed thousands of these systems and know what works with each setup. If the inlets and outlets are improperly placed, you might not be able to adequately access the water you are storing, and fixing the problem after your system is built could be expensive and time consuming.

 TIP! Even on small barrels, you will have to access the water by hose or bucket. Make sure the outlet spigot is high enough off the ground that you can place your receptacle underneath it without bumping the tap.

The size of the inlets and outlets are determined by the source of your water and the ways you plan to use it. For example, if you want to slowly irrigate your lawn with your water, you would not need a large outlet. Conversely, if you are storing water for emergency fire use, you will need a large outlet. Again, your contractor or manufacturer can help you determine the best size for your personal needs.

There are also a few optional features you can include with your inlets and outlets. Each tank manufacturer offers unique proprietary bells and whistles, so be sure to ask about these add-ons when you are ordering your tank. These devices offer a multitude of benefits but also will increase the cost of construction and add extra maintenance to the operation of your tank. Many of these can also be added at a later date if you feel they will improve your storage — so do not feel you have to buy everything at the time of initial construction. Just a few to consider include:

- **Inlet meters.** These are just like the meters used in regular household plumbing and will allow you to gauge how much water is going into your tank.

- **Inlet float valves.** This is similar to the ball used in a toilet tank and will shut off inlets when the float valve reaches a set point in your tank. This is for use in tanks where water might pour in quickly during a rainfall.

- **Aerators.** These are often integrated into the inlet and mix air into the water so the bacteria will be more likely to break down.

Overflow

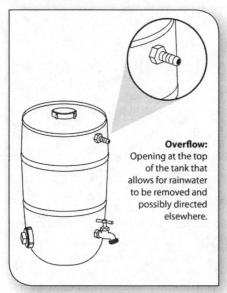

Overflow:
Opening at the top
of the tank that
allows for rainwater
to be removed and
possibly directed
elsewhere.

Most large tanks should have an additional overflow pipe that will help remove water if it begins to spill over the top. These extra features can direct the water out and away from your tank instead of having the water flow over the edges — or worse yet, push up against the cover and cause the roof to buckle or pop off your tank. The overflow also is a device for skimming dirty water off the top of the water level by allowing you to siphon off the layer where debris might be floating. There are many designs and versions and will often be included in your entire tank design or kit. The requirements needed for your overflow will depend on the tank capacity and incoming water source. If the water source is unpredictable, such as rainwater, and can quickly overfill your tank, having an overflow is the best insurance you can buy to protect your tank's integrity.

Service Access

Service Access: Large openings at bottom and possibly top of tank used to clean out and perform maintenance.

When choosing a site and planning your water tank, you need to consider its future use. If you need to gain access to the inside of the tank, fix a leak, or do other maintenance, you will most likely have to get inside the tank. For large tanks, this means you will need a trapdoor of some kind and a way to reach the top and bottom of your tank.

Choose a design that has an access point at the top so you will be able to reach all the areas where pipes enter into the tank. Most designs include a ladder integrated right into the structure itself — similar to a silo used on most farms or grain elevators. The best designs have ladders on the inside and outside so you can quickly get in or out of the structure without having to search around for a ladder. This is a simple safety issue because you do not want to be trapped inside a 20,000-gallon tank if water starts pouring in. For even more convenient access, consider selecting a tank with an additional ground-level access door for times when you need to remove built-up sediment from the bottom. With this handy feature, you can just open the access and scoop out the sludge to a waiting truck and not worry about hauling buckets up the ladder and out the top.

Consider access around the tank and the inlets and outlets, too, when setting and placing your holding tank. You might need access for a cement truck when you are building your tank, and future use might require driving in a truck to clear out built-up sludge. Make sure the path is large enough to accommodate these vehicles and the terrain is not too rough for them to get through.

CAUTION! When working in your tank, always follow the best safety practices. Do not work in a tank with water in it or with inlets turned on. Let someone else know when and how long you expect to be in the tank. It is best to have them watch from above while you work. Take a ladder with you inside the tank so you can quickly get out if needed.

Drains

The time might come when you will need to drain your tank. It might be accumulating too much sediment on the bottom and you will want to clean it out. You might have to move a smaller tank to a different location. The water might be contaminated and not usable. Whatever the reason, emptying out the tank completely through a separate drain — and not an existing outlet — is something to consider when you first install your tank. It is possible to release most of the water out through the regular pipe, but it will be difficult to completely drain the tank from this opening.

As you research tank design, look for those with built-in drains. Select a design that has an easy-to-use open drain mechanism but not so easily opened that it can be accidentally opened. As with inlets and outlets, the size of the drain depends on how much water you are storing. Ask the tank manufacturer or contractor for specifications on tank-to-drain recommendations.

> **⚠ CAUTION!** For safety, keep all openings of your drain covered when not in use so debris, critters, or even children's fingers do not become trapped in the drain hole.

If possible, place your drain at the lowest point of your installation, build the floor of your tank with a slight downward slope toward the drain, or install a "sump" around your drain. A sump is an area dug out around the drain so water and sediment will naturally flow and collect around the drain. These features will facilitate moving the water toward the drain.

The pipe leading out from your drain should extend far enough out from your tank so water will not accumulate around the edges. Make sure the end of the pipe is accessible and capped when not in use so critters do not crawl into the pipe toward your water supply. If possible, direct this drainpipe to an area where the water can be put to use and not wasted — it is also recommended to line the area immediately at the end of the pipe with gravel so it will not erode from the gushing water.

Bugs and Rodents

Any bit of stored water can become a potential watering hole for every crawling, flying, and climbing pest out there. Once a bug or animal falls into your water, it will die, and this can spoil the entire tank of water.

Avoid pests completely by keeping your water tank covered and screened. This includes the top and any openings that come into the tank, such as the inlet, outlet, and drainpipe. Seal the openings around each pipe, too, so the littlest bugs cannot get in. Critters might also consider your outlet pipes a place to seek shelter. Make sure to keep these tightly screened. You might even consider setting traps around the edge of the outlet. Also, keep thick grass or brush cut away from the exterior of your tank because small animals, such as rats, chipmunks, or squirrels, will avoid moving through areas without grassy cover.

Most insects who depend on water for laying their eggs also do not like moving water. If you live in an especially mosquito-infested area, install a bubbler or stir up the water every day or so. *If you are not planning to drink the water, you can also use Mosquito Dunks, as covered in the "Keeping the Water Clean" section in Chapter 4.*

Safety Features

In addition to the safety of the water stored in your tank, you must consider the actual integrity of your tank and the safety of those around it. The safety of the tank means that your tank is built to withstand the unique demands of your region. This means using materials or locating your tank so that it can stand up to shifting earth, freezing temperatures, hurricane winds, or whatever else your weather can throw at it. These times of emergency are when you will most need your water, so consider the "what-ifs" when you install a holding tank. Just a few examples of factors to think about would include:

- Using non-rigid, flexible piping in an area prone to earthquakes. A broken pipe would compromise your water supply, and flexible pipes are less likely to break when shifted around.

- Locating your holding tank so high winds from a tornado or hurricane cannot topple the heavy, water-laden tank.

- Choosing a material that can stand up to fire if you are planning on storing water for fire suppression. It will not work if the tanks melt and release the water before you are able to spray it across the flames.

- Taking measures, such as adding insulation or heaters, or draining the tank to protect against freezing. Simply stacking straw bales around the tank might be enough to keep the water inside from freezing.

Tank integrity also means to regularly inspect your tank for cracks, bowing, or buckling walls — these could indicate a tank collapse that would release water and be potentially dangerous to homes or people in the path of the water. Cracks in the wall of the tank could also allow contaminants into your stored water and make the water unsafe to drink. Also important is to choose a tank material designed to hold water that will not rust or corrode over time and release toxins into your water. The choices needed for your area are similar to those used for installing outdoor plumbing or waterlines. Talk with a local contractor or the staff at your local home improvement store to get ideas for what works best in your region.

As for the safety of people around the tank, this mostly applies to the issue of drowning. A large water tank holds as much water as a swimming pool and carries the same hazards. Make sure your tank is not accessible to children, especially if your tank is located in a remote area where curious children might be likely to explore. If possible, install a locking cover, a fence around your tank, or lock access to the ladder so children will not be tempted to climb to the top of the tank. Make sure your tank is also anchored in some way so it cannot tip over, which could injure people nearby and spill your collected water. Adequately secure any low-level valves or drains so people do not accidentally bump them open or release the water. For your safety, make sure the top access areas are non-slippery surfaces so you do not fall into the filled tank while you are working. If you are installing an underground cistern, make sure the ground above it is clearly marked or fenced off so you do not drive heavy equipment over the tank because this could crush the tank.

Aesthetics

Above-ground holding tanks are not the loveliest features to add to your landscape. Depending on your region, you can beautify your tank with trees, bushes, or even a brick wall. Make sure these trees are not planted too close to the tank because the root system can eventually grow into the footings. Also, if you live with any kind of homeowner's association requirements, make sure you check with the board before placing any type of tank. Some condo associations will not even allow a rain barrel.

Planting shade trees around your tank will also offer the added benefit of providing shade to your tank. This is a two-fold benefit because it blocks the sun from helping algae and bacteria grow, and it keeps the temperature of the water cool, which makes it more useful for drinking or plant watering.

Additional Maintenance Features

As with the enormous range of tank designs, there are also hundreds of bells and whistles you can add to your design. These features do not increase your water capacity but will improve the function of your storage and make maintenance easier. Ask your tank manufacturer or contractor for suggestions and additions available for your design. These also add more need for technical management and can add more cost to the budget. Research the many options online, or ask your tank sales staff for suggestions based on your specific needs. These are a few ideas to make water storage more manageable:

- **Alarms and Switches.** These types of devices can be set to automatically shut off power, close inlets and outlets, or set off an alarm based on your parameters. For example, you could choose a float switch that shuts the inlet when your tank is filled or an alarm to sound when your water level drops too low.

- **Exterior level indicators.** These have many different designs, depending on the type of tank you are using. An outside indicator is a simple addition that makes it easier to check the water amount in your tank without having to climb in.

- **Aerators and diffusers.** Aerators will keep the water infused with air, which will improve the taste and deter bacteria growth. Diffusers keep the incoming water from stirring up solids on the bottom of large tanks.

- ⊕ **Meters and gauges.** These run the gamut, and you can find a meter or gauge to measure everything from water temperature to line pressure to daily usage.

Calling in the Pros for Help

Even with all these extra features and diligent maintenance, the time might come when you need professional help. This will depend on your own level of fix-it skills, the gravity of the situation, and the amount of money you want to spend. You will know when you need professional help. It will most likely be times when you have a major leak and water is going everywhere you do not want it to go. It might be that your system is not keeping up with your supply or demand. You might find that you just are not able to manage the system you already have set up. Worst-case scenario: Your drinking water is making you sick, and you do not know why.

Obviously, your health is more important than anything, and if you feel stored or collected water is unsafe, ask for help. If you have worked with a contractor, a professional from the SWCD, or even a tank manufacturer, you most likely will have assistance readily available to you. If you have installed everything properly and followed the maintenance regimens, your tank and systems will be covered by a warranty. Call first before you try to fix big problems yourself because you could just end up making a bad situation even worse.

Using Your Water

Now that you have gathered and stored all this precious water, you will need a way to get it to its intended use. The best choice for delivery starts back at the plan you made for your stored water. What did you plan on doing with this water? Do you plan to use it for showering or just watering your trees and bushes? Will you take drinking water into the house to purify it? Once it is purified, where will you store that water? If you are going to use the stored water for irrigation, how will you get the water from the tanks to the areas to be irrigated? Are you storing water for an emergency, and, if so, how will you allocate the water? You also might need to retrain yourself and your family on ways to use water more efficiently — this can be done even if you are not storing water.

The following sections provide more information on the options available for pumping, techniques for efficiently using your water, and ideas for extending your water stores during an emergency.

This chapter will also cover the issues related to water storage and use for fire suppression. Many of the areas overlap, and you can pick and choose what works best for your family to create the most usable water storage plan.

Water Pumping Options — Electric, Wind, Hand, and Solar

A simple, small rain barrel only requires hooking up a standard garden hose and turning on the spigot. However, with large, more complicated systems, it will become difficult to manage the water without some kind of pumping delivery system or multi-tank storage system. In times of emergency, an old-fashioned rope and bucket will work for extracting water out of your storage tank, but for daily, long-term use, you will want a little more time-saving pumping power.

Pumps to move water have been around as long as water has been extracted from the ground. Hand pumps are still used at many primitive campgrounds across the country and require only human power to get the water moving. Drive along any country road, and you will eventually see an old windmill still churning away. It might not be hooked up to the well anymore, but you can see how this old-fashioned technology is still quite useful today.

These primitive pumping methods have gone through many upgrades over the years, and the price of these new devices can reflect this engineering. As with many systems related to water storage, your choices are wide-ranging and only limited by your determination and pocketbook. Additional choices in today's market include a wide selection of solar-powered pumps that need only bright sun to operate. Adding a pump to your system also adds extra maintenance tasks, as with any mechanical system. Consider, too, that an electrical pump should back up solar- or wind-powered pumps for days when there is no wind or sunshine. These naturally powered pumps, though, are essential for areas without access to electricity, such as back-country watering stations for livestock or for emergencies when fuel or electricity is unavailable to power your equipment.

With the broad range of pumps, it is recommended that you do more research beyond the information listed in the book. *Appendix A includes some resources to start with, and the Internet offers more current information. Search for "solar, wind, hand, electric, or siphon pump," and you will find many manufacturers.* Ask your local contractors, too, for tips on selecting the right pump for your application and region.

Do an investigation before purchasing a pumping system. Each brand offers a different advantage, and you will have to decide which feature works best for your situation. Choices range from automatic start buttons to power-saving shutoffs to long-lasting filters. Visit the many blogs now online for tips on what works for others in your area. You might even find that you could get by without a pump and save yourself time and money.

The size and type of pump you need is determined by the type of water coming into your tank and going through the machinery as it is pumped out. For instance, a gray water system will have some debris that could clog up the filters in an ordinary pump. For this reason, many people opt for a standard sump-pump, designed to handle less-than-clean water during floods. Another consideration when you select a pump is the distance you need to pump your water and whether you will need to pump this water uphill. With luck, you will be able to design your storage tank close to where you will use the water, but this is not always possible. Think about the route the water will take to its final destination. If you find potential distance or slope challenges, select a stronger pump that will be able to handle these demands. Product packaging on the pump will describe the capabilities of the pump, such as maximum pumping height or quantity. Choose a pump that exceeds your expected needs. This extra capacity might come in handy if you ever decide to add to your system. It is much easier to install a "too big" pump now than go back later and try to retrofit your tank.

 TIP! If you are storing water for emergencies, you might not have electricity to pump the water out. Have a backup pumping plan that will carry you through natural disasters.

Many pumps are easy for a do-it-yourselfer to install. Just open the box, set the pump in the bottom into your barrel, plug it in, and turn it on. Conversely, your pump could require an electrician, a plumber, and an inspector to get everything running. During your research, consider the installation requirements of your favorite pumps, especially as it relates to the added cost and labor. If you have many pipes coming in and out of your barrel and demanding pumping needs, it is best to check with an expert before investing in an expensive pump.

Pump filters

A necessary component of a pump is its filter. This little piece keeps the debris from getting into the motor, clogs up the works, and potentially ruins the equipment. Hand-in-hand with the filter is the disgusting task of cleaning the filter out to keep it running. You can potentially avoid using a filter if the water coming into your holding tank is pre-filtered through a screen, net, or even sand. *These methods were covered in previous chapters.* Removing debris from the water before it gets to your pump will save you the nasty job of fishing around in your tank and handling a gooey filter covered with who knows what. The type of filter you need will depend on your system and the pump you decide to purchase. Be sure to carefully read the manufacturer's instructions regarding cleaning and replacing filters.

Water Storage for Fire Suppression

In remote areas, people depend on their stored water for preventing the spread of fire to livestock and homes. This type of use requires a carefully planned storage system and pumping mechanism and is best approached by working with your local fire department or city officials. They will help you design a system and might even be able to offer some financial assistance with construction. It is to their benefit to prevent fires, too, and they are happy to help homeowners proactive enough to take this matter seriously.

> **CAUTION!** No animal, house, or personal property is worth dying for. This is the overriding message to consider if you are ever put in this situation. Do not risk your life if it looks like the fire is getting out of control, and always call the fire department first before you start fighting the fire yourself. Even if they are located a long distance away, they will eventually arrive with more firefighting power than you can provide on your own.

There are a few specific concerns related to firefighting with which the experts will help you. Specifically, they will address location and amount of stored water; pumping and delivery systems; and a use and exit plan for when fire is moving toward or taking over your property. Realize up front that it is difficult to store enough water to put out a large fire, but you can store enough water to keep the fire at bay until help arrives. Knowing the capacity of your tank and having a plan for how to use the water will help you determine where best to use this water in case of fire. Also, a central location or multiple locations will give you the best access to the water because you never know where a fire will break out. If you live in a dry area prone to wildfires, you might want to place your tank so you can create a protection perimeter around your home.

As mentioned previously, if you are storing water for fire-fighting purposes, the tank material must be able to withstand a fire — including the delivery pipes leading into and out of your holding tank. Choose steel or stone, or bury your tank underground where fire cannot get at it. If your water tank will be used for fire protection, it will be worthwhile to invest in a pump to help pressurize and move water out of the tank. Though, water supplied from a well might not be accessible if the power is destroyed due to the fire. A backup generator or eco-powered pump should be kept for this reason. In addition, do not forget all the other water sources around your home if fire strikes; swimming pools, animal watering tanks, and even water heaters can provide at least a dozen buckets of water. It also is insurance to have artificial fire suppressants on hand, such as fire extinguishers.

Separate Tanks for Purified Water

Keeping a separate tank for purified water is not a necessity, but it is a relatively inexpensive addition that will make your life run more smoothly. As mentioned in the purification section, most methods of "cleaning" water can take time — sometimes up to 24 hours. When you are in need of water, this can be a long wait. By keeping the purification process constantly going and consequently storing up the purified water, you will always have access to potable water. This can be accomplished by filling jugs or barrels as they are purified, but for larger amounts setting up a separate tank will be much simpler and more efficient. The basic setup for two or more tanks would include a tank for holding the water that needs to be purified and a tank that holds the clean water. This second tank has to be treated in the same manner as any other storage manner for clean water.

Additional Tools for Using Stored Water

▶ Portable solar-powered water heater

Beyond tanks, hoses, and pipes, there are a few handy items available on the market that make using your stored water simple and efficient, and most important, help you conserve water. Campers and hikers developed many of these devices, which are particularly useful in times of emergency. Modern technologies have made these devices even more portable and handy when power is not available. You will not incorporate these devices in your daily use, but keeping them on-hand for emergency is a great idea. Search online for "camping equipment" or "portable water gear," and you will find everything from siphons pumps for rain barrels to portable water heaters to solar showers to foldable 10,000-gallon water tanks. Your imagination and budget only limit the choices. *Appendix B also includes many resources for finding portable water devices.*

CASE STUDY: A SCHOOL COLLECTS RAINWATER FOR ATHLETIC FIELD IRRIGATION

Katherine Peragine
Catawba County Soil & Water
Conservation District
Environmental Educator
Newton, North Carolina

A high school in our county wanted to create a rainwater system that would allow them to divert rainwater from the school buildings into an underground cistern. This water would be used to irrigate the athletic fields. They wanted to do this to lower their water costs and to use a system that would conserve water during periods of drought.

A member of the Parent Teacher Association (PTA) from the high school contacted our office to see if the new athletic building would qualify for a cistern. We met with the PTA member, members of the school's administration, and maintenance workers to address concerns anyone might have.

Hickory High satellite view

▶ Satellite photo with the athletic building marked. (Catawba County Soil & Water Conservation District)

The one issue everyone had was safety. The idea of an above-ground cistern posed too many concerns to be feasible, so we looked at an underground system. The two downspouts in the following photo are on the west side of the athletic building, so we pursued the possibility of intercepting the storm water they discharged.

Downspouts

▶ Two downspouts considered part of the rainwater system. (Catawba County Soil & Water Conservation District)

The Hickory City Schools Administration Office supplied us with a drainage plan of the building, which showed the two downspouts were piped together into an 8-inch PVC drain tile and piped to a nearby storm drain.

Map of Drainage Plan of the School

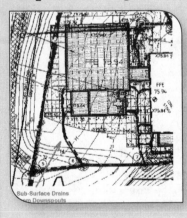

▶ Plan of the drainage system on the high school property. The arrow is pointing at where the main drainage line runs to a storm drain. (Catawba County Soil & Water Conservation District)

On this map of the drainage system, the downspouts are marked, followed by lines to represent underground pipes that lead to a storm water drain. This was already in place, and the goal was to divert this storm water before it reached the storm drain.

We met again to discuss the possibility of installing the cistern between the building and the storm drain. No one had any safety concerns, and everyone felt that this was the appropriate place to install the cistern.

Placement of the cistern

▶ The two people in the middle of the picture are standing near the storm drain. The plan was made to place the cistern underground between the storm drain and the sidewalk. (Catawba County Soil & Water Conservation District)

The parts were delivered, and the excavation began. The Cistern was delivered on a flatbed trailer. The ground near the drainpipe needed to be dug to connect the pipe to the cistern and to have a large enough hole to bury the cistern. The pipes and other materials were ordered and purchased at a local hardware store.

Cistern to be installed

▶ The large cistern was buried with the caps on top, which allows water to be pumped. (Catawba County Soil & Water Conservation District)

This cistern has to be completely buried under the ground and some water added to it to keep it from floating to the surface until the ground settles around it and encases it.

Digging commences

▶ The backhoe begins to dig a hole big enough to bury the cistern. (Catawba County Soil & Water Conservation District)

The most important lessons are learned from experience. During the planning phase for an underground water collection system, the landowner (a public school system) was advised to call about the location of underground utility lines before digging commenced. Their contribution to the project included time and labor during the construction process, so they would be responsible for the earthmovers and grading. The representatives felt confident their blueprints were accurate and digging commenced. They were wrong, and they end up rupturing a water main and cutting into a power cable. This delayed the progress because utility companies had come to fix the damage.

It is important to check and double check before you do something costly and unfixable. In the United States, 811 is a federally mandated phone number where citizens can get information about the location of utility lines before they dig. By calling this number, you can proceed with your underground project with safety and confidence. When you call the number, you will be directed to the local One Call Center and the affected utilities, employees which will then mark underground lines for free. More information can be found at **www.call811.com**, so you can be sure you "Know What's Below."

One obstacle we anticipated was how to attach a tee to the 8-inch drainpipe. A tee is a pipe shaped like a "T" that allows water to be split into two pipes. Some of the water was being diverted to the cistern, while the rest of the water or overflow from the cistern would continue to the storm drain. We looked at all the possible ways to make the connection and surmised that a saddle tee was the best option. It allowed us to make the connection without excavating a large area of the pipe. Only a short span was excavated, and a 4-inch hole was drilled into the bottom of the pipe, which allowed the saddle tee to be glued and clamped in place.

Saddle tee

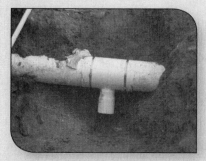

▶ Saddle tee is attached and glued into place. Only part of the original pipe needed to be cut away, and only a small amount of dirt needed to be removed from around the pipe. (Catawba County Soil & Water Conservation District)

We took elevation readings so we could verify the excavation depth. A 6-inch base of compacted sand was put in place for the cistern to sit on. The cistern was lifted into place, and the supply connections were made. The cistern was also filled about halfway with water prior to backfilling the hole; this prevents the cistern from floating out during the next rain event until the soil around the cistern settles.

Pipe Connected to Cistern

▶ The water is diverted from the saddle tee to the cistern. With everything in place, water is added to the cistern, and the backhoe fills in the hole (Catawba County Soil & Water Conservation District).

The hole was partially backfilled, and a foot valve, float switch, and vent pipe were installed. The foot valve, float switch, and valve are connectors that allow the water to be pumped from the cistern. It rained the day after the installation, and the water filled the cistern.

The focus of this project was to make the cistern as safe as possible to the students and the cistern. The chosen pump came with the pressure tank already mounted, which makes the system work like any home tap. Also, a float switch was mounted in the tank and wired to the pump, so when the cistern level dropped below a certain level, the pump would not come on. This prevents the pump from being burnt up and ensures that the pump did not lose prime. Prime is water needed in the pipe to allow the water to be pumped to the surface.

Diagram of the cistern system

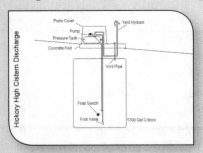

▸ This diagram shows all the components of the system that have been installed. (Catawba County Soil & Water Conservation District)

Buried cistern

▸ Small pipes were installed above ground to attach a well pump to extract water. (Catawba County Soil & Water Conservation District)

Holes were drilled into the equipment pad for the plumbing and electrical, and the pump was bolted down.

Well pump installation

▸ Pipes from the cistern were connected to the pump, and electrical connections were made. (Catawba County Soil & Water Conservation District)

Wiring diagram

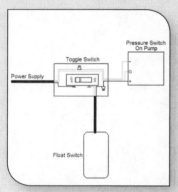

▸ Wiring Diagram of the Pump Switch. (Catawba County Soil & Water Conservation District)

The wiring diagram shows how the power supply goes through the toggle switch, which allows the entire unit to be turned off manually. From there, the power goes through the float switch, which only allows the pump to come on if there is sufficient water in the cistern. Last, the power goes through the pressure switch on the pump, which turns the pump on when the hydrant is opened and off when the hydrant is closed. With all the connections made and the power turned back on, the pump worked flawlessly.

Pump working

▸ Water from the pump flows, and installation is completed. (Catawba County Soil & Water Conservation District)

A fiberglass cover was bolted down to protect the pump from the weather.

With the cover in place and a lock on the hydrant, the safety of the students has been addressed. The pump is now ready to be used to water the athletic fields in the background once a hose is attached.

Well pump cover

▸ Fiberglass pump cover. (Catawba County Soil & Water Conservation District)

Water Usage in an Emergency

During an emergency, water conservation is not enough to protect your water supply. When water becomes scarce, you must conscientiously use every drop wisely and drastically change your daily water routine. A water emergency plan includes storing up water and outlines for how this water will be used and rationed during an emergency.

This usage plan will be different from a normal use, even for the most conservation-minded people. For example, in normal life, you might have started reducing shower times as a way to save water. This is great for conservation, but that four-minute shower can still use up to 10 gallons of water. With limited water access, you will have to adjust your family's daily patterns with the most important water needs in mind, which means secondary needs, such as laundry or showering, might have to be eliminated.

The dangers of dehydration

Most doctors agree that the human body can only survive for three to five days without water, so dehydration is a serious health concern. Dehydration occurs when the body does not get enough water to continue functioning normally. Dehydration can occur from not drinking enough fluids, but it can also result from illnesses that cause diarrhea or vomiting because of the rapid fluid loss involved. Unfortunately, these types of illnesses are most common during emergencies because of the lack of properly treated water and insufficient medical care. This is why being able to sufficiently purify and treat your water during an emergency is important.

People drink water and take in other fluids and water-dense foods, such as fruits, vegetables, and soups, to hydrate their bodies. For emergency-planning purposes, plan on having a minimum of 64 ounces, or 8 cups, of drinking water per day for each member of your household. Not becoming dehydrated in the first place is much easier than treating it, so rationing drinking water is not advised. In addition to 64 ounces of water per day, add low-sugar fruit juices, fruits with high water content, such as melons, or low-salt canned fruits and vegetables. Be sure to save and reuse any liquid in canned goods. Avoid drinking alcoholic or caffeinated beverages and eating salty foods, as these can lead to dehydration.

Some members of your household might be more prone to dehydration or might require more water than others. These groups include the young, the old, those with illnesses, pregnant or nursing mothers, and people who take medication. Also, if the temperatures are high, try to stay out of the sun during the hottest times of day to reduce the amount of fluids lost through sweating. Without proper hydration, it will not take long for a person to become dehydrated, and if it is not treated immediately, death is a real danger.

Emergency usage plan

Even if you do not expect the emergency situation to last long, begin using major water rationing right from the start. A few examples of ways to ration and conserve water during an emergency include:

⊕ **Showers:** Skip showers as much as possible. When you cannot stand sponge baths anymore, use a solar-shower shower bag. These bags contain a few gallons of water in a dark bag, which quickly heats up in the sun. Open the valve on the shower bag, wet your hair and body, and then close the water valve. Bag showers are available at camping stores.

⊕ **Baths:** Stick to sponge baths, and focus on washing your feet, hands, and face.

⊕ **Drinking water:** Do not ration drinking water unless you are in a desperate situation and running out. Drinking half of your recommended daily amount of water will still cause dehydration — it will just take longer to become ill. Try to use a water bottle with ounces measured on the side. Make sure each family member has an easily identifiable water bottle, and keep a chart that lists how many ounces each bottle is filled with each day. If it is hot or some family members are doing demanding physical work, more drinking water is required

⊕ **Cooking water**: Try to cook foods that do not require additional water, such as foods that need to be boiled or cleaned before eating. If you must cook food in water, add the remaining water to subsequent cooking, such as soups or pastas. During an emergency, flavor will not be your primary concern, but hydration will be.

- **Toilets**: Water that has been used for rinsing dishes or showering is not clean enough to drink, but it can be poured into the toilet tank to be used for flushing. Untreated water can also be used for flushing to save time and treatment supplies.

- **Tooth brushing**: Place a small amount of water in a cup. Dip the toothbrush into the water to wet it, and then put toothpaste onto the brush. Brush teeth normally, use half of the water in the cup to rinse your mouth and the other half to rinse the toothbrush.

- **Pets**: Measure out the daily water for your pet instead of pouring random amounts into the water bowl. Dogs will spill less from a water bowl that is less than half full. Do not be stingy when it comes to obtaining water for your pet, but determine the animal's daily need, and make sure you keep track of the amount the same way you are for other family members.

- **Cleaning:** During extended emergencies, you will have to eventually do some cleaning that requires water. Your cleaning needs include dishes and eating surfaces as the most important; towels and bedding next; then minimal laundry; and everything else comes last. Start your emergency storage with a supply of disposable plates and silverware, so you will not have to wash dishes right away. Also, stock up on disinfectant wipes, hand sanitizer liquid, and wet wipes. These work well for quick cleanups and do not require additional water usage.

The Three Bucket System

For these uses — especially cleaning or dishwashing — try using the "three-bucket" system to capture the gray water daily household activities create. An example of this would be for dishwashing: Bucket No. 1 holds clean water with dish soap in it; Bucket No. 2 holds clean rinse water; and Bucket No. 3 will eventually contain the dirty dish water from Bucket No. 1. After you are done with your washing, pour the dirty water from Bucket No. 1 into Bucket No. 3, and then use this water as needed to flush toilets or wash clothes. If this gray water is filled with food debris, it should not be used in the toilet because these chunks could clog the pipes. Pour the water through a strainer or coffee filter before filling the tank. You also skip filtering the water and use it directly on plants and landscaping. Bucket No. 2 can then be used to refill Bucket No. 1 for the next round of dirty dishes. The basic premise with this system is to always find a second use for water.

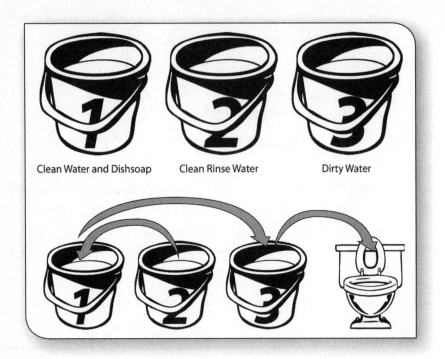

Clean Water and Dishsoap   Clean Rinse Water   Dirty Water

Surprise Emergency Water Sources in Your Home

If an emergency situation arises and you either do not have water stored or you begin to run out of stored water, there are some surprising places in your house to get the necessary water. If you think any sewer or water lines were damaged in the emergency, close off the main water valve to your house to protect the water you already have inside. Do this before attempting to use any of the water sources listed:

⊕ **The hot water heater.**
There should be a spigot or
valve at the bottom of the
hot water heater, which can
drain the tank. You must
make sure the electricity is
off at the main breaker, even if there is a power outage, or
the pilot light is off if you are going to drain the tank. Once
you have turned off the power source, simply drain the
tank into some buckets. A typical water heater tank holds
about 40 gallons when full, and the water cycles through
whenever a hot water faucet is turned on. This water will
be stale or flat, but it is still drinkable. *The next section offers
ideas on improving the taste of flat water.* Older water heaters
might have a little bit of built-up sediment at the bottom
— make sure you do not drain this into your buckets.

- Note: You will need to refill the heater before
  reconnecting it to the power or gas.

⊕ **The water pipes.** You can pull any water from your
cold water pipes by placing a bucket under the lowest
cold water faucet in your home and turning it on. Then
go to the highest cold-water faucet in your home and
open it to allow air into the system. This should send the
water in the cold water pipes within the house out of the
lowest faucet. Depending on the size of your house, you
could gain several gallons of water this way. If you live
in a single story home, make sure all faucets are turned
off, and place a bucket under the lowest faucet you can

find, such as the bathtub tap or where the water comes into your washing machine.

- **Waterbed.** If you have a waterbed in your home, the bed can be a source of water in an emergency. Some waterbeds contain toxic chemicals, so if you plan to use a waterbed as an emergency water source, you need to fill it once a year with water treated with bleach. If you do use the water in an emergency, drain the mattress and boil or treat the water with chlorine for drinking, or save the water for toilet flushing, shower bags, or washing clothes.

- **Ice.** If you have any ice in your freezer, pull the ice out and let the ice melt into drinking water. Do this early on, especially if you lose power because the ice will melt anyway. Also, consider storing filled milk jugs in your big deep freeze. This is handy in an emergency and a way to conserve energy year-round by keeping your freezer full.

- **Canned foods.** Many canned vegetables and some fruits are packed in water. Save that water when you cook the vegetables. Drinking green bean water might not be tasty, but in a true emergency, staying hydrated is more important than loving the taste of things.

- **Bathtubs, sinks, and baby swimming pools.** If you have time and know a water emergency is coming, fill your tubs, sinks, and baby-wading pools with water, and let it sit. You can always dip into these stores with a bucket for extra water.

Improving the taste of boiled and treated water

Boiled water or chemically treated water will not taste like the water you are used to drinking right out of the tap. Your treated water might have a lingering odor of chlorine, and freshly boiled water might taste stale, but this water is still perfectly safe to drink. It is just not palatable.

Using a mechanical or additional purification method, such as those mentioned in the previous chapters, will alleviate some of this problem, but there also a few quick fixes for emergencies that can help fix the taste issue. In an emergency situation and if you are thirsty enough, you will not care what your water tastes like. Keeping these handy supplies with your stored water, though, will help you through longer emergencies and provide creature comfort in a difficult situation.

After boiling, add oxygen by stirring the water with a sterilized spoon or pour it from one sanitized container to another. While the water is still hot from boiling, use it for herbal or decaffeinated tea, hot lemon water, or gelatin. For chemically treated water, the best remedy to poor taste is to also incorporate air. If chlorine has been used to treat the water, just let it sit in an open air container (protected from debris getting in) for 30 minutes. This will allow the fumes to dissipate a bit, which will improve the taste. Be careful, though, as FEMA guidelines state that properly disinfected water should have a slight bleach scent.

Flavored drink mixes are also another way to mask the taste of flat or funny-tasting water. Many brands are now available on the market that mix right into water without the addition of sugar. Some even offer extra electrolytes, which is nice to have on hand in case of dehydration. Avoid using high sugar content or caffeinated drink mixes because these can affect how your body absorbs water. These can be purchased in small and large quantities at any grocery store and will keep in a dry environment. Store these flavor packets and other drink mixes in a sealed plastic container with your emergency water supplies.

If you are trying to get children to drink the water, adding a small amount of fruit juice might make the water more appealing to them. You can also try adding fruit slices — lemon, lime, or orange — to add variety. Do not let citrus slices sit in water overnight though; the pith (the white spongy material on the rind) will make the water taste bitter.

Preparing meals with reduced water

Feeding your family during an emergency will require some creativity in acquiring supplies and choosing foods to cook. As part of an emergency plan, make sure to stock your pantry with food that can be cooked with little or no water. Most canned foods, such as vegetables, fruits, and soups, already contain liquid, so you will not have to add more, and you will gain a little hydration benefit from that food. For example, canned vegetables can be cooked in their own juice, and the juice from cans of fruit can be added to water for extra flavor. Cans of chicken or beef broth will also come in handy for cooking rice and other grains. Beware of the sodium content in these products, however, and choose no-salt or low-sodium for your emergency supplies. As mentioned previously, salt intake can lead to dehydration. Also, when washing produce or cooking with water, be sure to save this water and reuse it for dishwashing or toilet flushing.

 TIP! Stow away a can opener, matches, and cooking pots with your emergency pantry supplies.

Conservation Tips for Everyday Life

This book has offered up a broad range of ideas for using water wisely. Many of these ideas focused on rerouting water that has already been used or channeling water so it is maximized. You can, however, change your lifestyle in minor ways to conserve water right from the source. Teach yourself and your children to be conscious of how you use water in your everyday lives and to conserve as much as possible. Turn off the water while you brush your teeth, soap up your hands, shampoo your hair, and shave. Do not just run the water down the drain until it turns hot. Instead, keep a bucket in the sink to capture this water for other uses. Try adding a few of these ideas into your family's life, and they will soon become second nature.

In the kitchen

Start here by being aware of how you use — or misuse — the water from your tap and kitchen appliances. Consider these easy-to-try tips:

Drinking water. Instead of letting the water run until it is cold enough to drink, store a pitcher of water or filled plastic bottles in the fridge to conserve water. The water will already be cold when you want to drink it. There are certain types of water bottles available for purchase that do not leave an aftertaste, as some plastic bottles do. Look for bottles that are Bisphenol A (BPA)-free. BPA is a substance that is sometimes used to make types of plastics some countries, such as Canada, have labeled as a toxic substance. Many water bottle companies now use BPA-free plastics, so check the label for the BPA-free designation.

Cooking. You do many water-heavy things in the kitchen that you might not even think about, such as rinsing vegetables. You can fill a sink with clean water and rinse your vegetables and fruits at once rather than running the water.

Garbage Disposals. Disposal units might be convenient appliances, but they are water hogs. They use water to work properly. In addition, if you are using a septic system, that debris can fill up the tank quickly and cause problems. Consider starting a compost pile for all your organic trash. This will save water but and give you a source for high-quality garden soil.

Dishes. If you must do dishes by hand, avoid wasting water during the process. If you have a double basin sink, fill one side with soapy water, and use the other to rinse your dishes. Allow the dishes to soak, and scrub them before turning on the water to rinse them. If you have a single basin sink, you can fill a large basin or pot for soaking your dishes in hot water and use the sink for rinsing.

Dishwasher. Dishwashers are more water-efficient than hand-washing dishes — as long as you only run the dishwasher with a full load. Also, it is not necessary to "pre-wash" or rinse your dishes before putting them in the dishwasher. The technology now used in dishwasher detergents is designed so the soap is attracted to food particles, and your dishes will get cleaner if they go in "dirty."

In the laundry room

Even if you are not redirecting your laundry water to a gray water use, you can still save a significant amount of water in the laundry room.

Run a full load. The washing machine uses about 10 gallons to run a full cycle. Whenever possible, wash full loads of laundry, and avoid using the permanent press cycle because it uses an additional 5 gallons of water. If you cannot wash a full load, be sure to adjust the setting to "small" so your washer will adjust the water.

Do less laundry. Save water by reducing the amount of laundry you do each week. You do not have to wash something after one wearing. Consider going two days before washing slacks, towels, or blouses. For families, assign each member their own color of towel so you will know which one is yours, and hang it up after your shower so it will dry.

Upgrade or replace your washing machine. If you have an older washing machine, consider replacing it with a front-load, high-efficiency washing machine or Energy Star-rated appliance.

These newer models use less detergent and energy and as much as 50 percent less water than older washers.

In the bathroom

Water is used in the bathroom, and there are many ways to reduce your usage without any noticeable effect on your regular habits.

Toilets. Use the toilet only for traditional means. Do not flush trash or used cigarettes. You are wasting an average of 6 gallons a day and also creating more toxic waste. Use a trash bin for facial tissue, feminine hygiene products, or other small bits of trash. In addition to saving water, you reduce the likelihood of clogging your plumbing.

When to flush. One toilet flush uses almost 3 gallons of water. The old adage of "if it's yellow, let it mellow; if it's brown, flush it down" could save you water every day.

Low-flow showerheads. Install low-flow showerheads in all your showers. Most showers use 2.5 gallons of water or more per minute. Using a low-flow showerhead can reduce that number by nearly half without affecting the pressure you feel while showering. A showerhead converter is another device you can attach that will stop the flow once it gets warm and will start again after a pause. These devices are available at most hardware and home improvement stores. Installation is simple and involves just screwing the new showerhead on.

Faucet Aerators. Another easy installation to your faucets is an aerator, which works by forcing air into the water flow and helps maintain water pressure and increases rinsing ability. They restrict the flow, and some can reduce water usage by 32 percent without a noticeable difference. Aerators are inexpensive and can be purchased in most hardware stores and home improvement centers. These can also be installed on kitchen faucets.

Tank filler. Put bricks or plastic bottles filled with an inch of sand, marbles, or stones into the top water tank of your toilet. Screw the tops on the bottles, and place them in the tank. Make sure they are not near any of the moving parts or the hole. These bottles will displace the water, so it will take less water to refill the water tank after it is flushed. Do not use more than two 20-ounce bottles, as you will need enough water to flush the toilet properly, which is about 3 gallons. If there is less than that amount available, people might be inclined to flush twice, which can use more water.

Dual-Flush Toilets. Consider replacing your old toilets with newer models that advertise a "low flush" or include a "dual-flush" option. The low-flush toilets are designed to operate on 1 to 2 gallons per flush. The dual flush gives you the option of using different amounts of water per flush by holding down the handle longer. You can also install an adjustable toilet flapper that includes a dial on the flap to adjust the quantity of water used per flush. These are available at home improvement stores.

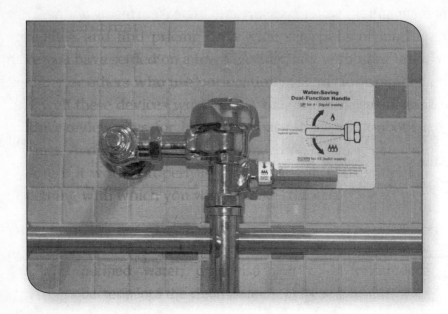

In your plumbing

You might be wasting water and not even know it. Check your plumbing systems for these hidden water drains:

Drips and Leaks. Make sure your pipes, toilets, and faucets do not leak. Even a small drip can waste about 20 gallons per day. How to tell if your toilet is leaking? Add a few drops of food color to your toilet tank where the water fills on top. Wait about an hour. If you see color in your toilet bowl, you might have a leak. Tank parts are inexpensive, easy to install, and a quick fix to a big problem. You can also check for leaks by turning off all your faucets and checking your meter. If the meter is still running, you might have a leak.

Insulation. Be sure your water pipes — in your basement and under your sinks — are insulated using pre-slit pipe insulation. This allows the water to quickly heat up and reduces the amount of water wasted during showers or bathing.

Outside

Much of the rainwater chapter focused on how to direct rainwater to your gardens. You can use that water most wisely by selective planting and conservation-minded lawn care. Here are just a few ideas for when you are planning your landscape:

Plant selection. When putting in new grass, shrubs, plants, or trees, choose varieties that are drought-resistant and appropriate to your climate. Do not try growing water-needy plants if your summers are hot and dry. Ask your local garden supply store for recommendations for varieties that will thrive in your zone.

Plant Groupings. Group plants together by their watering needs. Planning your garden in this way assists you with giving your plant the appropriate amount of water and allows you to maximize the efficiency of the water you are using.

Mulch. Use mulch around the base of your trees and plants to help retain water and reduce evaporation. Mulch is also useful in reducing weeds that steal water from your plants. Add a layer of mulch about 2 to 4 inches deep around the base of trees and plants.

Sprinklers. When installing sprinklers, make sure the water is aimed toward the grass and plants and not at the street or gutter. Do not water on a windy day because the water will evaporate before it reaches the plants.

Watering. Be sure you are not under-watering or over-watering your lawn. You can test whether your lawn needs water by stepping on it. If the grass springs up, the grass is not in need of more water. If the grass remains flat, it needs to be watered. Do not water too lightly, though because a light sprinkle will evaporate quickly and not go deep enough into the soil. This will encourage a shallow root system that is not water-efficient. Try placing an empty tuna or cat food can in the lawn as a test. If you have used enough water to fill the can, which should be about 1 inch, you have properly soaked your lawn.

Mowing. Let your grass grow to at least 3 inches before cutting it because this leads to greater water retention in the soil and will mean less watering to keep the grass healthy.

Timing. Water your lawn in the earliest part of the day. If you wait until nightfall, you could encourage the growth of fungus. If you water your grass during the hottest part of the day, the water will evaporate quickly and will not have enough time to soak into the roots, which means your lawn will require water more frequently.

Soaker hoses or drip irrigation. Instead of sprinklers, try using soaker hoses. These special hoses are laid flat on the ground in your garden. Instead of water only coming out of the end, there are small holes all along the length of the hose that allow water to be evenly distributed into the soil, which reduces evaporation. These hoses allow water to penetrate deep into the soil. Along with soaker hoses, also consider using a drip irrigation system. These systems are set up around your garden and allow water to drip slowly over a long time. They ensure greater absorption by the soil and less evaporation. You can buy these products at most garden supply stores.

Car Washing. Avoid washing your car at home. Many automatic car washes recycle water to conserve. If you are washing your car at home, do not leave the hose running while you are soaping up your car. Turn the hose on only to rinse your car, and buy a nozzle that can make rinsing more efficient. By leaving your hose running while it is not in use, you can waste up to 150 gallons of water. Also, consider using a waterless car wash system.

Conclusion

This book has walked you through a wide range of options in the world of water conservation, collection, and use. Some of these systems are as old as civilization, while others continue to evolve as technology and an innovative spirit rules. One thing that everyone agrees on is that saving water is one of today's most pressing issues. If current methods of storage and usage do not change, the water crisis will only get worse. Fortunately, you are now armed with the tools you need to truly remove yourself and your family from this cycle of misuse.

From rainwater harvesting to gray water reclamation to large-scale water retention, many of these practices have been around for generations. They have survived as preferred methods, because they are reliable and easy to implement. They have grown because conservation-minded people like you have kept them alive and updated them with modern technology. Water collection and storage is no longer a quaint, old-time practice but a modern-day strategy for survival. With the knowledge you now have, you can develop your own system that will take care of nearly every water need your family has.

Making even a small change, such as installing a rainwater barrel, can pay off in big rewards to your bottom line and to the Earth's underground water supplies. These small steps will lead to bigger changes in your lifestyle, in your children's outlooks, and possibly in your community's attitude toward water use. Your initiative in making a change can be a groundbreaker for those curious about this topic. Now that you have read the book, share with others how easy it is to use water efficiently and how satisfying it is to know you are not wasting this precious resource.

If you have decided to store water for emergency use, you also will feel the satisfaction of knowing your family is taken care of, no matter what Mother Nature has to throw at you. This "plan-ahead" approach is also another life lesson for your children. With any luck, you will never need to access your emergency stores. If the day comes, though, you will be prepared and can focus your energies on taking care of those around you.

If you have elected to take on the even bigger challenge of full gray water usage or large-capacity water collection, you are ahead of the game in the water resources world. Many people talk about it, some even plan a little, but most are deterred by the fear that it will be too hard to maintain or get used to. You know otherwise and, ideally, you will spread the word to all who are willing to listen. This sort of water "lifestyle" is critical to the future of our water supply and it is forward-thinking, take-charge people who can be the change in the world.

Thank you for taking this exploration. We hope you have found inspiration in these pages to make the leap into smarter water use and that you will search out more information and share what you have learned with your family, neighbors, and local officials. Your efforts will pay off in building a better future for everyone.

Bibliography

"Aquae Urbis Romae: Katherine Rinne on the Waters of Rome." EternallyCool.net. 04 March 2009. Web. 24 June 2010. **http://eternallycool.net/2009/03/ aquae-urbis-romae-katherine-rinne-on-the-waters-of-rome**.

"Aquifers, from USGS Water Science for Schools." U.S. Geological Survey, 29 Mar. 2010. Web. 21 June 2010. **http:// ga.water.usgs.gov/edu/earthgwaquifer.html**.

"Bermuda's Architecture," Bermuda-online. org. 21 June 2010. Web. 24 June 2010. **www.bermuda-online.org/architecture.htm**.

Ferrier, Catherine. "Bottled Water: Understanding a Social Phenomenon." April 2001. Web. 10 November 2010. **http://assets.panda.org/downloads/bottled_water.pdf**.

Black, David. *Living off the Grid.* Sky Horse Publishing: New York, 2008.

"Brazil: Rainwater harvesting in semi-arid region helps women," IRC. 14 Aug. 2008. Web. 24 June 2010. **www.irc.nl/page/42973**.

Burch, Jay D; Thomas, Karen E. (1998). Water Disinfection for Developing Countries and Potential for Solar Thermal Pasteurization. *Solar Energy*, 63 (1-3), 87-97.

Campbell, Stu. *The Home Water Supply: How to Find, Filter, Store and Conserve It.* Storey: North Adams, 1983.

Cavert, Katie (2010). Feasibility of Augmented Beeswax as an Appropriate Temperature Indicator for Solar Water Pasteurization Technology in the Developing World. Graduate Research Topic, Appalachian State University.

A Citizen's Guide to Bioremediation. Environmental Protection Agency. Web. 20 December 2010. **www.clu-in.org/ download/citizens/bioremediation.pdf**.

Clark, Josh. "What Is Gray Water, and Can It Solve the Global Water Crisis?" HowStuffWorks. com, 19 Nov. 2007. Web. 21 June 2010. **http://tlc.howstuffworks.com/home/gray-water.htm#**.

"Cleaning and Sanitizing With Bleach after an Emergency," Centers for Disease Control and Prevention. 15 Jan. 2010. Web. 24 June 2010. **www.bt.cdc.gov/disasters/bleach.asp**.

"Concrete Footings: Basics of Building a Concrete Footing," ConcreteNetwork.net. Web. 24 June 2010. **www.concretenetwork.com/concrete/footing_ fundamentals/introduction.htm**.

Curtis, Rick. *The Backpacker's Field Manual*, Random House Publishing:

"Dehydration," WebMD. 01 July 2009. Web. 24 June 2010. **www.webmd.com/fitness-exercise/tc/ dehydration-topic-overview**.

"The Effects of Bottled Water on the Environment" All About Water. Web. 10 November 2010. **www.allaboutwater.org/filtered-water.html**.

"Emergency Disinfection of Drinking Water," Environmental Protection Agency. 28 Nov. 2006. Web. 22 June 2010. **www.epa.gov/ogwdw000/faq/emerg.html**.

"Emergency Water Storage," Aquatechnology. net. 2001. Web. 24 June 2010. **www.aquatechnology.net/Emergency_Storage.html**.

"Examples of Rainwater Harvesting and Utilisation Around the World," United Nations Environment Programme. Web. 24 June 2010. **www.unep.or.jp/ietc/publications/urban/urbanenv-2/9.asp**.

Dowshen, Steven (MD). "Fluoride and Water," Nemours Foundation/Kid's Health. April 2011. **www.kidshealth.org**. May 22, 2011.

Grimes, Barbara. "Fact Sheet Grey Water," Web. 03 November 2010. **www.owasa.org/Documents/DocView.aspx?IDX=133.**

"FAQ from British Berkefeld on New Millennium Concepts," New Millennium Concepts, Ltd. Web. 22 June 2010. **www.britishberkefeld.com/faq.html**.

"FEMA: Water," Federal Emergency Management Agency, 4 June 2009. Web. 22 June 2010. **www.fema.gov/plan/prepare/water.shtm#1**.

GENERAL ASSEMBLY OF NORTH CAROLINA SESSION 2009 HOUSE DRH50427-MH-49. Web. 20 December 2010. **www.ncleg.net/Sessions/2009/Bills/House/PDF/H1385v0.pdf**.

Gould, John and Nissen-Petersen, Erik. *Rainwater Catchment Systems for Domestic Supply: Design, construction and implementation.* Intermediate Technology: Rugby, 2008.

Guru, Manjula V., Horne, James E. "The Ogallala Aquifer," The Kerr Center for Sustainable Agriculture, July 2000. **www. kerrcenter.com/publications/ogallala_aquifer.pdf**. 28 June 2011.

"Healthy Dogs Guide." WebMD, Web. 21 June 2010. **http://pets. webmd.com/dogs/guide/dog-dehydration-water-needs**.

"How Does Chlorine Bleach Work?" HowStuffWorks.com. Web. 24 June 2010. **www.howstuffworks.com/question189.htm**.

"How Does Chlorine Work to Clean Swimming Pools?" HowStuffWorks.com. Web. 24 June 2010. **http://science.howstuffworks.com/question652.htm**.

"How the AQUS Works," WaterSaver Technologies. 2009. Web. 23 June 2010. **www.watersavertech.com/AQUS-Diagram.html**.

"How to Read Your Water Meter," City of San Diego. Web. 21 June 2010. **www.sandiego.gov/water/rates/how.shtml**.

Jensen, James and Pavani Ram. "Hurricane Katrina: Health and Environmental Issues," MCEER, 17 May 2007. Web. 22 June 2010. **http://mceer.buffalo.edu/publications/ Katrina/07-SP02web.pdf**.

Johnston, Harold Whetstone. *Private Life of the Romans*. Scott, Foresman and Company: 1903.

Kinkade-Levario, Heather. *Design for Water: Rainwater Harvesting, Stormwater Catchment, and Alternate Water Reuse*. New Society: Gabriola Island, 2007.

Lancaster, Brad. *Rainwater Harvesting for Drylands and Beyond. Volume 1: Guiding Principles to Welcome Rain Into Your Life and Landscape.* Rainsource: Tucson, 2009.

Lawson, Sarah. "Virginia Rainwater Harvesting Manual," Cabell Brand Center, 2009. **www.cabellbrandcenter.org/Downloads/RWH_Manual2009.pdf**.

Ludwig, Art. *Create an Oasis with Greywater. Choosing, Building and Using Greywater Systems.* Oasis Design: Santa Barbara, 2009.

Ludwig, Art. *Water Storage: Tanks, Cisterns, Aquifers and Ponds for Domestic Supply, Fire and Emergency Use.* Oasis Design: Santa Barbara, 2009.

McManus, Barbara F. "Atrium," VRoma Project. Feb. 2007. Web. 23 June 2010. **www.vroma.org/~bmcmanus/atrium2.html**.

Miner, Dorothy L. "Emergency Drinking Water Supplies," North Carolina Cooperative Extension Service. Web. 24 June 2010. **www.bae.ncsu.edu/programs/extension/publicat/wqwm/emergwatersuppl.html**.

"National Hurricane Center," National Weather Service, 21 June 2010. Web. 21 June 2010. **www.nhc.noaa.gov**.

"National Rainwater and Greywater Initiative," Australian Department of Environment, Water, Heritage, and the Arts.

22 Feb. 2010. Web. 23 June 2010. **www.environment.gov.au/
water/policy-programs/nrgi/index.html**.

"NGO Forum for Drinking Water Supply and Sanitation," NGO
Forum. Web. 24 June 2010. **www.ngoforum-bd.org/projec-
tACaseControlStudy.htm**.

"NOVA Online: Roman Aqueduct Manual," Public Broadcasting
Service, Nov. 2000. Web. 23 June 2010. **www.pbs.org/wgbh/
nova/lostempires/roman/manual.html**.

Pogge, T. (2005). World Poverty and Human Rights. *Ethics &
International Affairs, 19*(1), 1-7.

"Rainwater Harvesting," TWAD Board. Web. 24 June 2010.
www.aboutrainwaterharvesting.com/rwh_methods.htm.

"Rainwater Harvesting: Germany," Rainwater-toolkit.net. Web.
24 June 2010. **www.rainwater-toolkit.net/index.php?id=21**.

"Rainwater Harvesting: Rainwater Basics," Texas A&M
System AgriLife Extension, Web. 22 June 2010. **http://
rainwaterharvesting.tamu.edu/rainwaterbasics.html**.

"Rainwater Tanks: Planning in South Australia," Government
of South Australia. 22 April 2009. Web. 23 June 2010.
www.planning.sa.gov.au/go/rainwater-tanks.

"River and Water Facts." National Wild and Scenic River System, 1 Jan. 2007. Web. 21 June 2010. **www.rivers.gov/waterfacts.html**.

"Sewer and Water Billing," Erie County Department of Environmental Services. 21 Jan. 2010. Web. 23 June 2010. **www.erie-county-ohio.net/does/billing/ws_rates.shtml**.

Smith, Kelly. "Pouring a Concrete Slab," I Can Fix Up My Home. 2008. Web. 09 March 2011. **www.icanfixupmyhome. com/Pouring_Concrete_Slab.html**.

Sobsey, M., Stauber, C., Casanova, L., Brown, J., & Elliott, M. (2008). Point of Use Household Drinking Water Filtration: A Practical, Effective Solution for Providing Sustained Access to Safe Drinking Water in the Developing World. *Environmental Science & Technology*, 42(12), 4261-4267.

"Solar Water Pasteurization," Solar Cookers International (2010). Web. 19 April 2011. **http://solarcookers.org/basics/water.html**.

Stanley, Doris. "Plants Clean Water … By Eating Fish Poop," National Science Teachers Association. 2001. Web. 23 June 2010. **www.ars.usda.gov/is/kids/water/story1/ fishwasteframe.htm**.

"Sun Cooking USA: Water Pasteurication Indicator (WAPI)" Sun Oven (2010). **www.sunoven.com/cart/index. php?main_page=product_info&products_id=8**.

"Survey: Memphians pay lowest water, waste water rates," Memphis Business Journal. 19 Feb. 2009. Web. 23 June 2010. **www.bizjournals.com/memphis/stories/2009/02/16/daily35. html**.

"Third Report" Centers for Disease Control. Web. 10 November 2010. **www.cdc.gov/exposurereport/pdf/thirdreport.pdf**.

Thompson, Andrea. "Study Reveals Top 10 Wettest Cities," LiveScience. 18 May 2007. Web. 24 June 2010. **www. livescience.com/environment/070518_rainy_cities.html**.

Tigno, Cezar. "Country Water Action: Thailand — Promoting Rainwater Harvesting, Preserving Rainwater Jar Culture," Asian Development Bank. Dec. 2007. Web. 24 June 2010. **www.adb.org/water/actions/THA/Jar-Culture.asp**.

"U.S. Indoor Water Use." U.S. Environmental Protection Agency, 12 May 2010. Web. 21 June 2010. **www.epa.gov/watersense/pubs/indoor.html**.

"US Water Supply," U.S. Environmental Protection Agency. 12 May 2010. Web. 23 June 2010. **www.epa.gov/watersense/pubs/supply.html**.

Walker, Brian. "Study: Millions in Bangladesh exposed to arsenic in drinking water," CNN. 21 June 2010. Web. 23 June 2010. **www.cnn.com/2010/WORLD/asiapcf/06/20/ bangladesh.arsenic.poisoning/index.html**.

Warwick H. and A. Doig. (2004). *Smoke – the Killer in the kitchen. Indoor air pollution in developing countries.* ITDG Publishing. London, UK.

Waskom, R. and J. Kallenberger. "Graywater Reuse and Rainwater Harvesting," Colorado Water Institute, Colorado State University. July 2009. Web. 22 June 2010. **www.ext.colostate.edu/pubs/natres/06702.html**.

"Water Efficiency," Environmental Protection Agency, 12 May 2010. Web. 22 June 2010. **www.epa.gov/watersense/water_efficiency/index.html**.

"Water Gardening," University of Illinois Extension, Web. 22 June 2010. **http://urbanext.illinois.edu/watergarden/about.html**.

World Health Organization (2004). *The Global burden of disease: 2004 update.* **www.who.int/healthinfo/global_burden_disease/2004_report_update/en/index.html**.

Glossary

Adjustable toilet flapper: You can install this device in your toilet tank to reduce the amount of water used for each flush.

Aquifer: An aquifer is a layer of underground, porous rock, gravel, sand, or clay that filters groundwater and transmits it below the groundwater level. A common practice for drilling a well is to drill the well shaft into an aquifer so the well is supplied with a constant source of water.

Artesian wells: The ancient Romans found a well near Artesium, which is now called Artois, in a province in France. The water flowed heavily from it, because the well had tapped into an underground aquifer. Now, the term Artesian Well refers to any well dug deep enough to access an aquifer.

Black water: Waste water from toilets. This water cannot be reused in households.

Cleanout: A cleanout is a special access in a pipe in which a pipe may be inspected for clogs and cleaned out using a plumber's snake.

Culvert: A large pipe made out of metal, concrete, or other impermeable material used to channel water flow. Frequently used under roads, bridges, or in ditches.

Drainfield: Called an absorption field or leach field, an open area of soil or grass that contaminated water or gray water is dispersed as a way to filter it, collect solids, or reduce contaminants.

Drip irrigation system: Systems set up around your garden that allow water to drip slowly over a longer period of time to water your plants. It ensures greater absorption by the soil and less evaporation.

Gray water: Waste water from household uses, such as washing machines and showers. This can also be written as graywater, grey water, or greywater.

Groundwater: Water below the soil surface that is the supply for natural springs and wells.

Fall: The vertical distance between a source of gray water and the destination of that water, usually outside for irrigation purposes or for interior plumbing.

Fog and dew collectors: Simple units that allow water to condense on screens, and the water is collected at the bottom of the unit. These can be freestanding and mobile.

Inlet: The portion of a water tank where it comes in.

Low-flow showerhead: A showerhead that can reduce the amount of water by 2.5 gallons of water per minute.

Oocyst: A spore or cyst, which contains a developing bacteria. They can survive for a long time without a host and are often found in unclean water.

Outlet: The portion of a water tank that allows water to come out. The outlet might be a simple spigot placed low on the tank or a more complex system involving a pump to remove water from the tank.

Plumber's snake: A long spring-like cable that is twisted and pushed through a pipe in order to grab and pull out clogs.

Potable water: Clean and safe for drinking, potable water is a term that can be interchanged with the phrase, "drinking water."

Hydrologic cycle: Water from the oceans and large bodies of fresh water evaporate into the air and condense to form clouds. If enough water condenses, it becomes too heavy and falls to the ground in the form of rain, snow, or other precipitation. Water that does not stay in the groundwater level runs into lakes and streams and eventually evaporates and becomes rain again.

Runoff: Water that cannot be absorbed by the surface it is flowing across.

Runoff coefficient: A number that represents how much water the particular material does not absorb.

Surge tank: A tank designed to slow down the flow of water as it surges into the tank. It is used to hold water to cool it down but not designed for long-term storage.

Water table: The space where water is held underground in soil and aquifers. The top of the saturation level is called the water table. In places where the water table intersects with a hill, it is common to find a spring or seep of water.

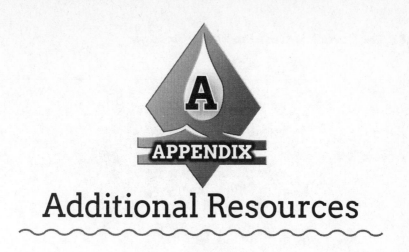

Additional Resources

This section will give you a starting point to find tank suppliers, design ideas, websites, and other important information as you design your own water storage system. You can also find a multitude of sources by searching online for "water storage," "rainwater," "gray water," or "water harvesting." The most reliable websites end in .gov, .edu, and .org. Be wary of individual blogs or commercial sites, because they might not be completely upfront about what they are selling.

USDA Water Quality Information Center at **http://wqic.nal.usda. gov/nal_display/index.php?info_center=7&tax_level=1&tax_ subject=596.** This site provides information about water availability and water quality around the United States. All of the information is free.

Solar Showers

Safety Central

http://safetycentral.com — This online company offers many different emergency preparedness products, including solar showers and different water containers.

Energy Supermarket

http://shop.solardirect.com — This site offers energy-saving devices. They specialize in solar energy.

Go Westy

www.gowesty.com — This is a camping supply site that has a wide range of outdoor products you might need with your water collection systems. They even have complete outdoor shower systems.

Low-flow Showerheads

Moen

www.moen.com

Dual Shower Head

www.dualshowerhead.com

Showerhead Store

www.showerheadstore.com

Faucet Aerator/ Low-flow Aerator

Eartheasy

http://eartheasy.com

Grant Water Innovations
www.bigdropletscattering.com/Faucet.html

Do it Best
www.doitbest.com

Adjustable Toilet Flappers

Water Miser
www.watermiser.com

Hardware and Tools
www.hardwareandtools.com

Tool District
www.tooldistrict.com

Water-Efficient Dishwashers/ Clothes Washers

Water, Inc.
www.waterinc.com

Energy Star
www.energystar.gov

Consumer Reports
http://web.consumerreports.org

Electrolux
www.electroluxappliances.com

Composters

Green Culture
www.composters.com

Nature Mill
www.naturemill.com

Clean Air Gardening
www.cleanairgardening.com

BPA-free Water Bottles

Nalgene
http://nalgene.com

Klean Kanteen
www.kleankanteen.com

REI
www.rei.com

Faucet/Pitcher Water Filters

Aquasauna
www.Aquasauna.com

PUR
www.purwaterfilter.com

Brita
www.brita.com

Drought-Resistant Grass

High Country Gardens
www.highcountrygardens.com

Lawn Care
www.lawncare.net/drought-resistant-grass

Warner Brothers Seed Company
www.wbseedco.com

Soaker Hose/Drip Irrigation

Best Nest
www.bestnest.com

The Drip Store
www.dripirrigation.com

Irrigation Direct
www.irrigationdirect.com

Waterless Car Washing Systems

Croft Gate USA
http://croftgateusa.com

Eco Touch
http://ecotouch.net

Water Storage Containers

Be Prepared
http://beprepared.com

Plastic-Mart
www.plastic-mart.com

Safety Central
http://safetycentral.com/water.html

The Epicenter
www.theepicenter.com

Pumps

Backwoods Solar Pumps (solar)
www.backwoodssolar.com

Grand Canyon Pump and Supply (solar and wind)
www.gcpsolar.com

Grainger (well and submersibles)
www.grainger.com

Aluminum Water Bottles

SIGG USA
http://mysigg.com

Gaiam
http://life.gaiam.com

Rain Barrels

Gardener's Supply Company
www.gardeners.com

Rain Barrel Source
www.rainbarrelsource.com

Simply Rain Barrels
www.simplyrainbarrels.com

Clean Air Gardening
www.cleanairgardening.com

Rain Barrel USA
www.rainbarrelusa.com

Rooftop-Cleaning Diverters and First-Flush Systems

The Gardener's Supply Company
www.gardeners.com

The Water Filtration Company
www.waterfiltrationcompany.com
(800) 733-6953

Gray Water System Suppliers

Dripworks
www.dripworksusa.com — This site offers drip irrigation plumbing supplies that can be used in gray water systems.

Formulas and Conversion Tables

The following formulas will help you convert rainfall into gallons. There are also numerous online calculators that will do the math for you. Visit **www.calctool.org,** and you will find calculators you can plug your criteria into.

Formula to convert inches to feet

In order to determine the amount of actual rainfall that falls on your entire catchment areas, you will first need to convert your average amount of rainfall from inches to feet. This conversion step will make future calculations easier to determine. Use the following formula

> Average rainfall per month in inches ÷ 12 =
> Average rainfall per month in feet

For example:

> 3.6 inches ÷ 12 = 0.3 feet of average rainfall per month

Formula to calculate square footage of catchment area

Lot length × Lot width = Lot square footage

For example, suppose you have a 90-by-50-foot lot:

90 feet × 50 feet = 4,500 square feet

(If you want to just determine what the catchment of your roof is, you need only use the dimensions of your home.)

Formula to calculate how much rain is being caught by your catchment area.

[Catchment area in square feet] × [inches of rain/12]
× 7.48 (to convert to gallons) = Actual amount
of rain falling on the catchment surface.

(Assume at least 25 percent of this sum will be lost to evaporation if surface is impermeable, and even more is lost if surface is permeable, such as grass.)

Liquid Conversion Table

1 Tbsp.	½ fl. oz.
2 Tbsps.	1 fl. oz.
¼ cup	2 fl. oz.
1/3 cup	2 2/3 fl. oz.
½ cup	4 fl. oz.
2/3 cup	5 1/3 fl. oz.
¾ cup	6 fl. oz.
7/8 cup	7 fl. oz.
1 cup	8 fl. oz./ ½ pint
2 cups	16 fl. oz./ 1 pint
4 cups,	32 fl. oz.
1 pint	16 fl. oz./ 1 pint
2 pints	32 fl. oz.
8 pints	1 gallon/ 128 fl. oz.
4 quarts	1 gallon/ 128 fl. oz.
1 liter	1.057 quarts
128 fl. oz.	1 gallon

1/3 Tbsp.	5 ml
3 tsps.	15 ml, 15cc
1/8 cup, 6 tsps.	30 ml, 30cc
4 Tbsps.	59 ml
5 Tbsps. & 1 tsp.	79 ml
8 Tbsps.	118 ml
10 Tbsps. & 2 tsps.	158 ml
12 Tbsps.	177 ml
14 Tbsps.	207 ml
16 Tbsps.	237 ml
32 Tbsps.	473 ml
1 quart	946 ml, 0.946 liters
4 quarts	3785 ml, 3.78 liters
1 gallon	3785 ml, 3.78 liters

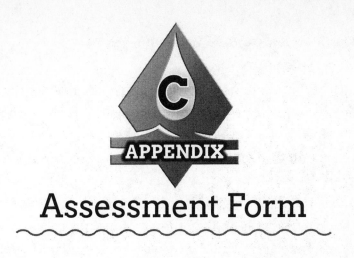

APPENDIX

Assessment Form

The purpose of this drawing and form is to make an assessment of your site, your goals, and your needs when you install a rainwater or gray water system. The site plan directions included here combined with your answers from the assessment form will help you plan out your system down to the last detail. Refer back to the main book for specifics on the issues listed on the form.

How to Draw a Site Plan to Scale

Many programs are available online to translate your measurements and plan an actual scale drawing. You can also use grid paper to map out the location of pipes, outlets, and storage tanks on your property. A place to find a basic sketch to work from is the appraisal of your house before you bought it. It will have all the square footage and placement of the structures on your property to refer to. Include the direction water flows on your property and any significant boundaries you need to consider. Also, if you are working with a professional designer or SWCD expert, he or she will provide a survey and site map for you at no charge.

To draw this yourself, begin by creating a key that works best for you. For example, one grid square equals 1 foot or one square equals 10 feet. It does not matter what key you use as long as you stay consistent. Collect all the measurements required to map out your plan. Depending on the type of system you are installing, these measurements might include your property line, incoming water sources, your home's square footage, pipes in your home, and areas where you want the water to be stored and used. Draw these all out on the grid paper as they relate to one another. Consider using colored pencil for each feature, such as blue for sewer lines and red for gray water outlets. Clearly mark each line as you draw it out so you will remember later on.

Once you have the general layout mapped out, try visualizing this to its true scale. Look over your plan, and imagine how you will access your storage tank and how you will connect up all the gray water lines. Once your plan is finalized, you can submit this to the permitting office (if needed) or give it to your contractor to work with.

Assessment Form

Many of the assessment issues you need to consider can pertain to gray water and rainwater harvesting systems. Use this form to assess both, and feel free to skip questions that do not apply to your situation. *Refer back to the list of questions covered in Chapters 4 and 5 and other issues discussed throughout the book.* If you are doing a gray water and a rainwater system, delineate the differences as you enter information, or copy this form and make two separate assessments by water source.

Laws, permits, and codes

Local Laws Found: _____

_____

_____

_____

_____

_____

Permit Needs: _____

_____

_____

_____

_____

_____

_____

_____

_____

_____

Official Contacts and Phone Numbers: _____

_____

_____

_____

_____

_____

_____

_____

_____

_____

Neighbors' concerns? How have I addressed them? _____

_____

_____

_____

_____

_____

_____

_____

_____

Tax incentives or cost-share availability? What are deadlines or requirements for participation? _____

_____

_____

_____

_____

_____

_____

_____

_____

_____

Contact info for these programs: _____

_____

_____

_____

_____

General

Goals (include phrases, such as more accessible water, easier access, and financial savings): _____

_____

_____

_____

_____

_____

_____

_____

_____

_____

_____

Lifestyle changes I am willing to make: _____

_____

_____

_____

_____

_____

_____

_____

_____

_____

_____

Specific system requirements

(Mark these plans on your site map.)

Where will I put this system — especially holding tanks? Will there be enough space? _____

_____

_____

_____

_____

_____

_____

_____

_____

_____

_____

Where will I collect the water? Specify identified plumbing issues, such as gravity-fed or watershed area: _____

_____

_____

_____

_____

_____

_____

_____

_____

_____

_____

How much can I expect to collect? Calculations from formulas in individual chapters and Appendix B. List collection by specific source (such as rain barrel for NE downspout): _____

_____

_____

_____

_____

_____

_____

_____

What will the final purpose of this water be? _____

_____

_____

_____

_____

_____

_____

Any issues that will affect my construction or installation (existing plumbing, landscaping issues, seasonal requirements, and soil permeability): _____

_____

_____

_____

_____

_____

_____

How much do I want to spend to build and maintain this system?
How much will I save or even gain financially? _____

_____

_____

_____

_____

_____

_____

Materials choices — List by pipes, pumps, filters, and tanks:

_____

_____

_____

_____

_____

**Purification plan — List goals and intended methods of
purification:** _____

_____

_____

_____

_____

_____

_____

What is my usage plan? Specifically list any issues related to emergency use, such as daily use plan, location of emergency supplies, contingency plan for power outage or having to leave your property: _____

_____

_____

_____

_____

_____

_____

_____

_____

_____

_____

Any other issues I need to address? _____

_____

_____

_____

_____

_____

_____

_____

_____

_____

_____

Project Plans

As mentioned throughout this book, many of these construction projects require specific skill and experience to be done right. If done improperly, your structure could fail, causing injury, death, water contamination, or financial loss. In some cases, a large failure could lead to breaking of local laws and a loss of access to water. Please consult with an expert and review the resource materials listed in this book for more information.

Plan No. 1: Step-by-Step Instructions for Building a Simple Foundation

Here is a simple way to pour a foundation for barrel, tank, or smaller-sized cistern (400 gallons or fewer.). If you have a larger cistern, you will need a professional builder to build a proper foundation.

Tools:

- ❑ Shovel
- ❑ 4-foot level or laser level
- ❑ Side-cutting pliers
- ❑ Hammer

Supplies:

- ❑ Wooden stakes
- ❑ Mason's string
- ❑ 1-inch-by-4-inch lumber for concrete form
- ❑ 2-inch-by-4-inch stud or aluminum screed
- ❑ Steel wire mesh for smaller projects, rebar for larger projects
- ❑ Tie wire
- ❑ Vapor barrier material (This may be needed for building code. Check with your local permit office to see if it is required before you begin.)

Instructions:

1. First, determine the location and the size of the foundation.

2. Remove grass, tree roots, big rocks, or clumps of soil from inside the roped-off area.

3. Flatten the dirt with your shovel and check it with your level.

4. Create a trench on the perimeter of the area. Trench width and depth depend on your climate and local codes. Footing trenches are 16 inches wide and 9 inches deep.

5. Optional step. You might be required to put a vapor barrier, which can be plastic sheeting on top of the dirt, before your pour the slab. To do this, place the vapor barrier over the dirt and allow it to hang over the perimeter.

6. If you are only going to put a rain barrel on top of the foundation, you can just use steel mesh, but if it is going to be a larger cistern, you will need to use rebar. Consult local building codes to help you decide which you will need. If you are using the steel mesh, you will use tie wire and connect the mesh at the intersections between the pieces of mesh. You will twist the ties with pliers. The mesh helps strengthen the footing where the pressure is the greatest from the weight of the rain barrel.

7. Once the steel mesh is laid out in the form, you will need to build the concrete form. This is a wooden frame that keeps the concrete in place while it is drying. You will use 1-inch-by-4-inch boards.  Use the string lines you created earlier to make sure your form is square.

8. Lay out the boards and cut them to desired sizes. To help keep it up and straight, drive more stakes on the outside of the form about every 16 inches. This will make sure the frame stays in place when you pour the concrete, because the weight of it will put pressure outward on the frame. Use the level to make sure the frame is level and square all the way around, and make adjustments if necessary.

9. Nail the frame together and nail the stakes firmly to the frame to make sure it is firmly together and supported.

10. After the frame is in place, it is time to prepare the cement. The best way to handle this for this size of a project is to rent a small cement mixer from a rental facility. You can get a small portable mixer for the day. If you are pouring a slab for a larger cistern, you might need a cement truck to come and pour the concrete instead.

11. Once the cement is mixed according to the manufacturer's instructions (make sure the correct amount of water is added, and check the consistency of the cement), point the chute into the frame. As the cement begins to fill the form,

use a shovel to push it around evenly — be sure to work in sections you can finish before the cement hardens.

12. After pouring the cement, use an aluminum screed or a piece of 2-inch-by-4-inch wood to level off the cement. Lay the board across your forms and work it across the surface in a sawing motion. Continue filling and finishing sections until the entire surface is filled.

13. Let your concrete cure for the full manufacturer's recommended time. Avoid walking on the cement, and keep it protected from heavy rain or hot sun. Check your permit before you place your tank.

Plan No. 2: A Laundry Drum Gray Water Surge Tank

To build your own system, assemble these tools and supplies and follow the simple directions listed here.

Tools:

- ❑ Handsaw
- ❑ Pliers
- ❑ Drill with a drill bit that can make at least a 1-inch hole in the wall of the washroom where the washer resides
- ❑ Flat-head screwdriver
- ❑ Caulk gun

Supplies:

- ❑ Threaded ¾-inch coupling
- ❑ ¾-inch bulkhead fitting and gasket
- ❑ Teflon tape
- ❑ Silicone sealant
- ❑ ¾-inch garden hose adapter
- ❑ ¾-inch hose (or the length you would prefer; this is to water the plants)

Optional supplies:

- ❑ 1-inch, 3-way diverter valve
- ❑ Hose clamp
- ❑ Barb to pipe thread adapter
- ❑ Two 90-degree bends
- ❑ 1-inch PVC pipe
- ❑ Stand pipe
- ❑ P-trap

Approximate cost:

- ❑ Barrel – used or refurbished: $50; new: $120
- ❑ Plumbing Parts – $14
- ❑ Other supplies – $10
- ❑ Total cost – $74 to $144

Instructions:

1. Decide how you want the water to exit the washer. If there is a window available, you can hang the washing machine drain hose out the window over the barrel. If this is not possible, you might need to drill a small hole in the wall for the drain hose to exit. When drilling a hole in the wall, choose a place away from outlets or pipes coming out of the wall. For extra safety, turn off the electricity and water while you are working. Be sure the hole is above the lid of the washer. This prevents water from draining from the wash unintentionally.

2. Hang the drain hose above the barrel. The barrel should have a hole on top for the water to flow into. You might want to secure a lid and cut a hole in the top or put mesh, such as chicken wire, over the top to prevent animals and debris, such as leaves, from getting in. The purpose of the tank referred to as a surge tank is to slow down the flow of the water and allow it to cool down before using it. It is not meant as a method of long-term storage for gray water.

3. Raise the barrel up off the ground. The simplest way to do this is to stack cement blocks or even bricks underneath the barrel. This allows the water to flow at a higher pressure from the hose you will install in the next step to be used for watering plants. Make sure the barrel is level and stable.

4. If a hole does not exist in the bottom of the barrel, you will need to drill one. The size of the hole will depend on whether the barrel has an open top or the top is sealed, because a sealed top will mean you cannot attach anything on the inside of the barrel. Below are directions for both types of barrels.

   a) Sealed Top Drum – Drill a 7/8-inch hole in the side of the barrel about half an inch from the bottom. Use a hammer and tap in the ¾-inch hose adapter until you reach the nut that is on it about half of the adapters length. Seal around the adapter with the silicone sealant.

   b) Open Top Drum – Drill a 1-inch hole in the side of the barrel about half an inch from the bottom. Push the ¾-inch PVC threaded coupling from the inside. It will have a nut and rubber gasket on the end. Wrap the threaded coupling sticking out through the hole with Teflon tape. Screw the hose adaptor over the tape for a tight seal. Once the hose adapter is installed, seal around the hole with silicone sealant.

5. The gray water system should be operational. Make sure the hose you attach to the hose adapter is long enough to reach the plants you want to water. The farther away from the tank you are watering, the less water pressure you will have. You can buy a small water pump and attach it to the hose. You will then attach another hose to the other end of the pump to water your plants.

6. Seal up the opening around where the hose comes out of your house. Use steel wool, spray foam, and caulk to seal it tightly so rodents do not find a way into your home.

Alternate Steps: You might want to have a three-way diverter that will allow the water to be switched from the gray water system to the septic or sewer system. You will need your optional supplies for these steps.

1. Attach the washing machine drain hose to the bottom of the T-shaped three-way diverter valve using a hose clamp. To one side of the diverter, attach a 90-degree bend to a small, angled elbow pipe.

2. Place the PVC pipe in the existing standpipe if one exists. This is the open pipe your washing machine drain was most likely draining into. If one does not exist, you will have to attach a standpipe, a PVC pipe, or a P-trap to the drain in the wall where the washing machine pipe was

draining into. Your 3-way diverter must be higher than the top of the washer, which is why the standpipe and PVC pipe might be necessary. You might need to use a handsaw to cut the PVC to the right height. Place the smaller end of the PVC pipe into the standpipe and the other end into the 90-degree angle attached to the 3-way diverter. The standpipe is then secured to the P-trap with tighteners you screw. If you want an idea of what the setup looks like, you can look at the pipes under your kitchen sink — they are set up in this manner — or refer to the diagram below.

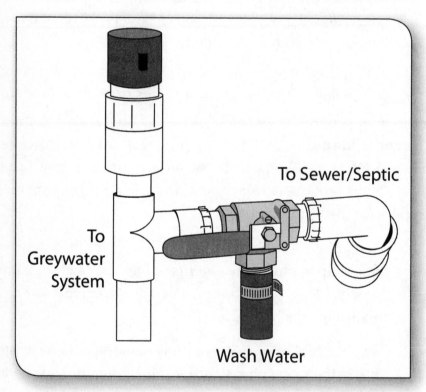

To Sewer/Septic

To Greywater System

Wash Water

3. On the other side of the diverter, attach another 90-degree
    bend. To the end of 90-degree bend, attach enough pipe
    to go through the hole in the wall to the outside. You can
    then attach one more 90-degree bend that points down into
    the top of the barrel. You want there to be space between
    the barrel and the pipe to prevent the water from backing
    up once the barrel is full or to put back pressure on the
    washing machine's pump.

Plan No. 3: Tankless Laundry to Landscape System

The laundry to landscape system is similar to the laundry drum system, but it does not use a surge drum, and instead, allows the water from the washing machine to go directly to the plants of your choice. This is an even more cost-efficient system, because it does not require the purchase of the barrel, and it is as easy to install as the drum system. In this system, you are using the force of the washer drain pump to push the water outside into multiple hoses to specific areas in your yard. It is more automated, because it sends water to the plants and areas you choose without the need to move any hose. One caveat is this type of system can overly stress the pump in your washer if the water will be pumped for a long distance to the area you want to water.

Tools:

- ❑ Handsaw
- ❑ Pliers
- ❑ Drill with a drill bit that can make at least a 1-inch hole in the wall of the washroom where the washer resides
- ❑ Flat head screwdriver
- ❑ Caulk gun
- ❑ Shovel

Supplies:

- ❑ Threaded ¾-inch coupling
- ❑ ¾-inch bulkhead fitting and gasket
- ❑ Teflon tape
- ❑ Silicone sealant
- ❑ ¾-inch garden hose adapter
- ❑ A number of ¾-inch hoses to the length you prefer. You can choose how many places you want the water to go.
- ❑ Two 1-inch 3-way diverter valves
- ❑ Hose clamp
- ❑ Barb to pipe thread 1½-inch adapter
- ❑ 90-degree bends (Figure out the number of outlets in your system to determine how many you will need.)
- ❑ 1-inch PVC pipe (Measure the system first to determine how much PVC you will need.)
- ❑ 1-inch PVC tee
- ❑ 1½- by 1-inch Bushing
- ❑ 1-inch PVC slip to barbed
- ❑ 1-inch slip to barbed tees*
- ❑ 1-inch HDPE hose *
- ❑ 1-inch barbed tees*
- ❑ ½-inch "green back" barbed ball valves *
- ❑ Standpipe
- ❑ P-trap
- ❑ Level

❑ Autovent

* Figure out the number of outlets in your system to determine how many bends you will need.

Approximate Cost:

❑ Plumbing Parts – $50 to $75

❑ Other supplies – $10

❑ Total cost – $60 to $85. The specialty plumbing parts can be purchased at a local plumbing store, or you can order them online at Dripworks (www.dripworksusa.com).

Instructions:

1. You will need a 3-way diverter in this system to allow the water to be switched from the gray water system to the septic or sewer system.

2. Attach the washing machine drain hose to the bottom of the T-shaped 3-way diverter valve using a hose clamp. To one side of the diverter, attach a 90-degree bend.

3. Place PVC pipe in the existing standpipe if one exists. This is the open pipe your washing machine drain was most likely draining into. If one does not exist, attach a standpipe, a PVC pipe, or a P-trap to the drain in the wall where the washing machine pipe was draining into. Your 3-way diverter must be higher than the top of the washer, which is why the standpipe and PVC pipe may be necessary. You might need to use a handsaw to cut the

PVC to the right height. You need to place the smaller end of the PVC pipe into the standpipe and the other end into the 90-degree angle attached to the 3-way diverter. The standpipe is then secured to the P-trap with tighteners. If you want an idea of what the setup looks like, look at the pipes under your kitchen sink or refer to the diagram in the Laundry Drum system section.

4. On the other side of the diverter, attach another 90-degree bend. To the end of 90-degree bend, attach enough pipe to go through the hole in the wall to the outside.

5. Attach another 3-way diverter to the pipe coming through the wall to the outside. On the opposite side, attach in this order:

    • 1-inch PVC Tee

    • 1½- by 1-inch Bushing

    • 1½-inch female adapter

    • Autovent

    This allows the line to be vented and prevents a vacuum from forming.

6. On the bottom side of the PVC Tee, add another length of 1-inch PVC pipe. It needs to be long enough to reach the ground. You may even choose to bury it and run the pipes under the surface of the soil at a depth of about 9 inches.

7. Connect a 90-degree bend to the 1-inch PVC pipe by connecting a 1-inch PVC Slip to barbed. Then add the HDPE hose.

8. Determine how many outlets you want and then figure how much PVC tubing, HDPE hose, 1-inch barbed Tees, and ball valves needed for each outlets. You can see an example in the diagram. Lay out the pipes where you want them. This will help you visualize how it will all fit together and how much pipe, hose, and other supplies you will need. Do not seal the pipes together until you are sure of your design.

9. Once you are sure of the design, dig trenches to place them in. Trenches must be at least 9 inches deep. If your ground has plenty of slope toward your gray water destination, you can lay the pipe in the ground. The ground needs to be at least a 2-percent slope or you will need to take extra measures to provide proper fall. These measures will be explained in the next step. If the slope is steep enough, put the pipes together and add silicone sealant to glue the joints.

10. *If the ground has plenty of slope, you can skip this step.* If there is not much natural fall in the land between your house and the destination of your gray water, create a slope of at least 2 percent as you are laying your pipes. If you are unsure, use a level, a small instrument that allows you to determine whether your pipes are vertical or are at an angle (or slope). Starting nearest to the house, place the pipes in your trench and use the level to check for slope. If the fall greater than 2 percent (or a slope of two-feet for every 100-feet), build up the soil in the trench to minimize the fall. Continue laying pipe and leveling until you reach the final destination. If the fall is not deep enough, dig

deeper on the far end of the pipe and add a little dirt on the nearer end, the end closer to the house, until it reaches the angle you desire. You do not want to go too far beyond 2-percent slope in any one section of the pipe or you might find yourself digging deeper. Assemble all of the pipes, but do not glue them yet. Check the slope to make sure it is as even as it can be through the pipe system. Once you are sure the slope is correct, glue the pipes together with silicone seal.

Placing valves, as shown, allows you to cut the flow of water to certain areas. Do not to shut off all lines or the back pressure will cause too much strain on your appliance's pumping mechanism.

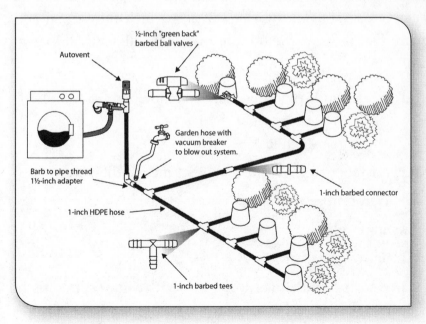

Plan No. 4: Making a Small Rain Barrel

This is an easy and inexpensive project to do in an afternoon. Make sure to place your rain barrel far enough away from your home so it will not cause water damage to your foundation if the barrel overflows.

Supplies:

A plastic barrel, trash can, or other similar clean container. Make sure the container you choose is solid with no hairline cracks or holes. Choose a size that can handle your expected runoff — *as shown in the formula in Chapter 3*; a one-inch rainfall on a 1000 square-foot roof can yield more than 600 gallons if all the water is collected.

- ❑ A hose and spigot mechanism made up of:
    - ❑ A ¾-inch spigot
    - ❑ A ¾-inch lock
- ❑ A rubber ring (washer) that is 1 inch in diameter
- ❑ Water pipe tape
- ❑ Super glue
- ❑ Silicone sealant
- ❑ A ¾-inch brass overflow valve with a corresponding hose adapter kit
- ❑ A screen to be used as a filter on top of the barrel

Instructions:

1. Check your equipment and gutter system for flaws or weaknesses before you begin. Make sure your gutters can handle the additional volume of water if you are rerouting the water flow, otherwise the water will just flow over the edge of the gutter and be wasted before it gets to your rain barrel. Replace the gutters with larger capacity pieces where needed.

2. To make your own barrel, drill a ¾-inch hole near the bottom of the barrel to affix your spigot system. Use plumber's tape to create a tight seal and silicone sealant to affix the spigot to the barrel with no leakage. Allow sealant to completely dry before going on to the next steps.

3. Your barrel needs a lid or screen to keep out debris. If your container already has a lid, drill or cut a hole in the lid large enough to fit your incoming pipe or downspout. If your container is not lidded, cover the entire barrel with a screen to catch the debris as the water flows in. With work gloves on, cut the screen to overlap the opening at the top of the barrel and attach the screen securely to the top of the barrel using nylon straps. Make sure the lid you use is removable so you are able to clean out your rain barrel when it gets dirty. You can also build a small roof over the top of your barrel — just make sure you do not impede the flow of rainwater.

4. After your rain barrel is assembled, give it a test run with your garden hose to make sure it catches, holds, and

dispenses water as you want it to. Make adjustments now before you put it in place.

5. Choose a location that is optimal for ease of collection and usage. Mount your barrel off the ground, such as on a pallet, stacked bricks, or a raised stand. This will allow you to use gravity during collection and usage of stored rainwater. Rain barrels are easily integrated into buildings with existing gutter systems that can be diverted directly into your rain barrel. Rain should run from the highest point on the structure into gutters that slope downward toward the rain barrel and allow the water to flow naturally in that direction for containment.

6. Divert or adjust the angle of your gutters so water runs directly into the barrel through the screen, or attach a flexible tube to the end of your downspout and direct this into the hole you have cut into the lid of your rain barrel. If your barrel's capacity is large enough and the downspouts will reach, attach two or more pipes into one tank. Otherwise place tanks at each downspout location.

Troubleshooting your rain barrel will be limited to tightening or waterproofing seals and increasing or decreasing the volume of water the system is capable of maintaining. If you find you are not collecting as much water as you can or need to, evaluate the efficiency of your guttering system to divert water toward your rain barrel. Simply raising the far end of a gutter by as little as one inch higher than the end nearest your barrel will divert enough water to fill your barrel easily every time it rains.

If you find you are having difficulty getting the water out of the barrel once it has been collected, you might need to consider raising it off the ground further to use a gravity feed. This will create more fall. Be sure you are not trying to move the water too far away from the rain barrel, or consider attaching an electric pump to spray the water in the location you would like to use it. *Pumping options were covered in Chapter 8.*

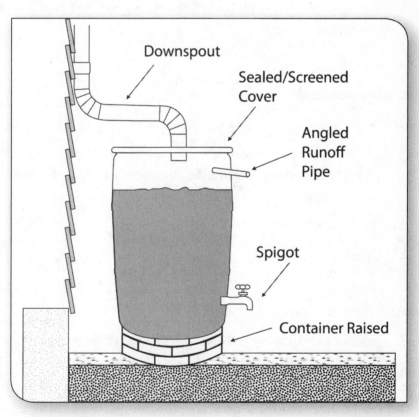

Plan No. 5: Earthworks Designed to Water Landscaping

Whenever you create a new landscaping area in your yard, such as flowerbeds, bushes, or trees, consider the structure of the land around it, and use this structure to naturally direct rainwater to these plantings. Water will soak into the soil and into the mulch. Even after the rain stops, water will continue to filter into the soil and water the tree.

Supplies needed:

❑ Shovel

❑ Gravel, quality black dirt, old timbers, or any other decorative element you would like to see in your landscaping

❑ Bark mulch

❑ Grass seed

Instructions:

1. Select a location at the base of a natural slope or depression. When planted, the tree or bush will sit slightly below the ground around it. When deciding how deep to dig the hole, take into consideration the recommended planting depth plus 2 to 3 inches of mulch, plus this slight depression.

2. Dig the recommended depth hole for your plant and place the plant in the hole.

3. Mulch around the new planting but do not place edging around the tree or bush.

4. Use the materials you have collected to create low berms that will naturally funnel rainfall across the surface of the earth to this depression crated around your new plant.

5. Seed with grass areas where you have dug that are not covered with mulch and water thoroughly.

Plan No. 6: A Direct to Landscape Pipeline

It is highly recommended that you consult with or hire a plumber for this project, because you will be cutting into the drainpipe that may reside in a wall. You can still advise or work with the plumber to explain what you want him or her to do for you.

Tools:

❏ Hacksaw

❏ Drill

❏ Caulk gun

Supplies

❏ 1-inch three-way diverter

❏ 90-degree bends (the number will depend on the slope of the ground and the direction you want the gray water to go)

❏ 1-inch PVC Pipe (the amount will depend on how far you want to divert the water)

❏ Silicone sealant

Instructions:

1. Determine where the drainpipe is that connects to the sink you wish to use gray water from. The pipe you will want

to access will be above any cleanout access and below any P-trap under the sink.

2. Cut a small section of pipe. Drill a hole through the wall behind the pipe so it lines up with the hole in the pipe and leads to the outside. The hole needs to be large enough to allow a 1-inch PVC pipe to be connected. Install the 3-way diverter between the two ends of the pipe you have created.

3. Attach a PVC pipe to the diverter through the hole you have drilled. Seal around the hole with silicone sealant or you might create an opening for rodents or insects to get into your house.

4. Attach a 90-degree bend facing the ground.

5. Dig a 9-inch deep trench from where the gray water will exit to the plants or mulch basin where you want the gray water to flow to.

6. Attach a length of PVC pipe to the 90-degree bend. It needs to be long enough to reach the bottom of the trench.

7. Attach a 90-degree bend to the bottom of that PVC pipe so the open end is facing down the trench toward its destination.

8. Attach a length of PVC pipe to the open end of the 90-degree bend. Make sure it is long enough to reach the plants or mulch basin. You might want to add silicone sealant to all the joints to make sure they are secure.

9. Make sure there is enough slope in the pipe as it moves away from the source. You can check it with a level. If the slope is 2 percent to 8 percent, you can safely bury the pipe. If not, refer to the soil sloping instructions in the Laundry to Landscape system above.

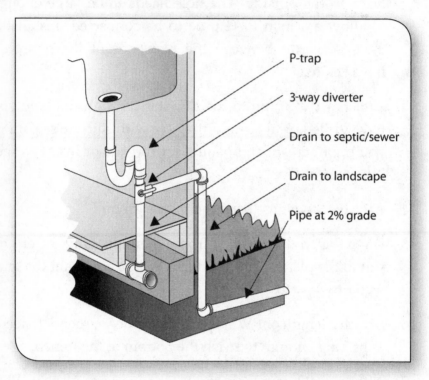

P-trap

3-way diverter

Drain to septic/sewer

Drain to landscape

Pipe at 2% grade

About the Author

Julie Fryer has written numerous pieces focused on green living and organic gardening. Her published credits include: articles in *Gardening How-To* magazine, contributions to **www.myorganicgardeningblog.com**, and the book *The Complete Guide to Your New Root Cellar*. She lives with her husband and two sons in southeastern Minnesota where they spend their free time fishing, camping, gardening, and enjoying the outdoors.

Index